MW00414559

# THE WATERLESS SEA

# THE WATERLESS SEA

## A Curious History of Mirages

CHRISTOPHER PINNEY

REAKTION BOOKS

*For Trudi Binns*

Published by
REAKTION BOOKS LTD
Unit 32, Waterside
44–48 Wharf Road
London N1 7UX, UK

www.reaktionbooks.co.uk

First published 2018


Printed and bound in China by 1010 Printing International Ltd

A catalogue record for this book is available from the British Library

ISBN 978 1 78023 932 3

# CONTENTS

# Prologue: Chasing Mirage

> Magic is never destroyed – the most we can do is to cut ourselves off, amputate the mysterious antennae which serve to connect us with forces beyond our power of understanding.
>
> Henry Miller, *The Colossus of Maroussi*[1]

In the 1870s, the naturalist Alfred Russel Wallace wrote to the prestigious science journal *Nature* enquiring whether any readers had encountered a 'good and honestly constructed cyclopedia', by which he meant 'one that does not send you hunting for information from one volume to another and refer you backwards and forwards to articles that do not exist'. Wallace then highlighted what he described as a particularly outrageous example of this 'will-o'-the-wisp', the elusive phenomenon of mirage. He describes his hunt for reliable information on 'unusual atmospheric refraction'. The entry on 'Refraction' directed him to 'Mirage, Fata Morgana', which in turn directed him to 'Refraction, Atmospheric, Extraordinary'. Returning to the letter R, he 'found the article "Reflection and Refraction"', but was here referred to '"Light, Optics, Refraction, Refrangibility"; then to letter A, "Atmosphere, Atmospheric" – nothing on the subject. Letter E, "Extraordinary Refraction" – nothing but a reference again to "Mirage"!'[2]

Wallace's misleading 'cyclopaedias' deserve to be as celebrated as Borges's Chinese encyclopaedia for their arbitrary taxonomy. Wallace's vision is the opposite of Borges's Aleph (the space which contains all other spaces) or Borges's imperial cartographers who constructed maps that duplicated and covered the territories they described. Wallace's whimsical cyclopaedia is most like a catoptric cistula, or theatre, a device for the endless magnification of objects, except that in Wallace's case it is an *empty* cistula. It is also of course a witty allegory of the nature of mirages, anomalous shape-shifting phenomena, forever mobile and mutable and which long escaped what Borges termed the 'rigour of science'.[3] My task in this short account is to populate the catoptric box, to provide locations and points of reference for the will-o'-the-wisp that so frustrated Wallace, without falling for the eccentric typology or overambitious cartography that preoccupied Borges. This account attempts to navigate a sinuous pathway through a mysterious and evanescent terrain, showing how mirages were yoked to other concerns, how they came to have a politics, and how we can continue to learn from their sublimity. It attempts to insert mirages into a world history, connecting them in patterns that exceed the narrow national concerns of nature writing or travel writing, revealing the ways in which mirages became prisms through which religions and systems of rule were imagined, and the very limits of vision were confronted.

I first became aware of mirages through a dangerous thing called Miraj. I was sitting in the front room of a Jain friend's village home in Malwa, central India. He runs a small shop supplying villagers with daily necessities such as matches, mouth-freshener and light bulbs. Customers would frequently call, often to buy small packets of chewing tobacco called Miraj, their brilliant yellow packets defaced with lurid images of cancerous mouths, as required by Government of India health guidelines. This dual nature of *miraj* as both poison and cure can serve as a leitmotif for the mirages I will consider in this book. These too will be incarnated as objects of enchantment, like Miraj's shiny yellow packaging, and also of vengeance, like the ulcerous palates which blighted those yellow packets.

1 James Abbott, *Mirage of a City Hidden in the Convexity of the Earth*, 1854, lithograph by T. Black at Asiatic Litho Press, after a sketch by Abbott. Abbott encountered the 'fairy structure' on which this image was based near Mhow in central India in 1829. Lithography was the perfect medium for the amplification of what, experientially, may have been less ornate.

ONE

# Strange Visions Under a Cliff in Central India, October 1829

Fortuitously, one of the most enchanting accounts of mirages was produced, 180 years earlier, very near to where I contemplated the poison and cure of Miraj tobacco. Major James Abbott was traversing the plains of central India in October 1829, en route to the town of Mhow in Malwa, when he first encountered a mirage of distant cliffs, usually seen shortly after sunrise. These cliffs were 'of so substantial an appearance[,] so marked with rent and fissures, so tufted with bushes, shrubs and lichens; so clear and distinct of outline, that it is scarcely possible for an unpracticed eye to doubt their reality'.[1] He promised himself several times that 'in the afternoon [he] would pay those cliffs a visit' but by then they had always disappeared, leading him to question 'whether [he] had not been suffering some illusion of the eye or mind'.[2]

These illusions (which were fancifully lithographed for a report in the *Journal of the Asiatic Society of Bengal* in 1854) were most definitely pleasurable: cliffs that laboured under a 'monotonous aspect' were 'diversified by and enlivened by the presence of a white town or of moving objects' and 'clothed with the richest verdure, stolen from green corn fields drawn up aloft as by enchantment to garnish the fairy structure'.[3] Villages

> buried beneath the convexity of the earth's surface are seen hanging reversed in the air, and should any small river with its boats be flowing there, all the shifting scenery would be presented in the clouds: the white sails, greatly magnified, and distorted, having a truly spectral appearance as they hover silently by.[4] (illus. 1)

Abbott is enthused by what he calls 'truly magical' optical effects and adds imaginary mirage-experiences to his catalogue of actual experiences.[5] It would be 'easy' to imagine, he writes, the 'glorious apparition which such a city as London would present reflected in mirage', and he then reproduces as a plate a beguiling and beautiful image of 'A Lake City as it Might Appear in Mirage', a panoramic image of a fantastical Oriental metropolis of minarets and domes suspended at the far edge of a serene expanse of water, in the foreground of which stand gothic ruins.[6] One cannot help seeing in this image Abbott's imagination of the Rajasthani city of Udaipur, elevated to similar 'magical effect' through the romantic writings of Colonel James Tod in his *Annals and Antiquities of Rajasthan*, the first volume of which was published in 1829, the year in which Abbott locates his earlier experiences.

Indeed, Tod himself had written at length about mirages, in an equally enthusiastic tone, in his personal narrative at the end of volume I of his *Annals* and it is almost inconceivable that Abbott would not have been familiar with this. Tod notes that on a cold winter's morning after leaving Mairta en route to Ajmer, he witnessed a 'magnificent *mirage*', and then proceeds to provide invaluable evidence of local terms for the phenomenon, the Maroo terming it *see-kote* (literally 'the cold weather castles', which he glosses as 'castles in the air'). Conversely,

2 James Abbott, *A Lake City as it Might Appear in Mirage*, 1854, lithograph by T. Black at Asiatic Litho Press, after a sketch by Abbott.

> in the deep desert to the westward, the herdsmen and travellers . . . style it *chittram*, 'the picture', while about the plains of the Chumbul and Jumna they term it *dessasur*, 'the omen of the quarter'. This optical deception has been noticed from the remotest times.[7]

Tod then suggests that *chittram* are inferior mirages, whereas *see-kote* are superior Fata Morganas (a complex form of superior mirage with elaborate horizontal and vertical features), in which the traveller might imagine that he would find 'a night's lodging' but would not 'expect to slake his thirst there'.[8]

Tod wittily alludes here to the Fata Morgana's often elaborate architectural forms, frequently imagined as palaces or cities in the sky where one could indeed obtain 'a night's lodging'. And he distinguishes these from what in northern India were known as *chittram*, common illusions of the type which appeared in the form of lakes where one might indeed 'slake [one's] thirst'. Modern physics explains the 'complicated temperature distributions, each of which has its own peculiar optical consequences'.[9]

Both kinds of mirage are created by the bending (refraction) of light as it moves through different temperatures and hence densities of air. In inferior ('lower') mirages such as the appearance of a lake in the desert, light passes more quickly through the unusually hot and hence less dense air near the surface of the desert. It bends upwards towards the spectator, who consequently experiences light from real objects, such as the sky, not above, but below their level of vision. Metalled

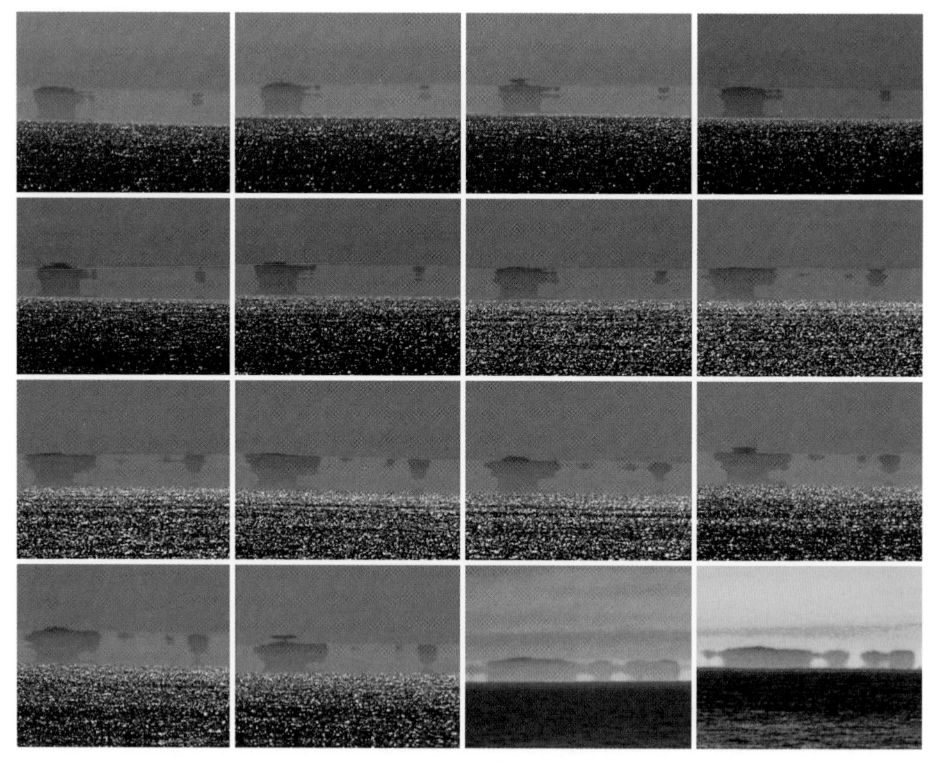

3 Sequence of images taken over several hours showing a Fata Morgana over the Farallon Islands near San Francisco.

highways provide the ideal environment as a thin layer of air becomes super-heated above the road. The ripples of the 'lake' and the shimmering of heat haze are produced by the volatility of the heat gradient and the consequent effervescent refraction. Inferior mirages are so-called because the mirage appears below the real object (illus. 3).

In superior mirages, by contrast, the mirage appears above the real object because of thermal inversions, in which a layer of warmer air sits above a band of cooler air. This explains their frequency in polar regions, especially over large sheets of ice and where the layers of air tend to be relatively stable. Refraction bends light upwards, often producing mirages of what appear to be castles and palaces (or other places where one could spend the night) whose real object may be hidden beyond the curvature of the earth. An early account of Fata Morgana in *Scientific American* drew on the work of the Jesuit physicist Joseph Maria Pernter in a very clear explanation of its optics. The beholder views the world through air whose density, especially in temperate regions, is continually changing. Pernter suggested that this was similar to the effect of viewing the world through 'inferior window glass full of "bubbles"'. As the viewer moves the glass from side to side, the world is seen through different thicknesses of glass and 'objects appear and disappear'. If the glass is of poor quality, there will also be a multiplication of images, as is the case with Fata Morgana:[10]

> attracted by a lofty opaque wall of lurid smoke . . . By slow degrees the dense mass became more transparent, and assumed a reflecting or refracting power: shrubs were magnified into trees, the dwarf *khyre* appeared ten times larger than the gigantic *amli* of the forest. A ray of light suddenly broke the line of continuity of this yet smoky barrier; and, as if touched by the enchanter's wand, castles, towers and

> trees, were seen in an aggregated cluster, partly obscured by magnificent foliage. Every accession of light produced a change in the *chittram*, which from the dense wall that it first exhibited, had now faded into thin transparent film, broken into a thousand masses, each mass being a huge lens; until at length the too vivid power of the sun dissolved the vision; castles, towers, and foliage, melted, like the enchantment of Prospero, into 'thin air'.[11]

A digression by Tod on Isaiah's Old Testament prophecy about the 'parched ground' becoming a 'pool' prompts interesting observations about the relation between the Arabic term for 'desert' – *sehara* – and *sehrab*, the term that 'inhabitants of the Arabian and Persian deserts apply to this optical phenomenon', which translates as 'water of the desert'. Tod then moves to the Roman historian Quintus Curtius' description of the Sogdian desert, in which 'there arises such an exhalation that the plains wear the appearance of a vast and deep sea'.[12] This idea of mirage as an 'exhalation' is something that we will encounter again in Chinese and Japanese accounts.

Tod's and Abbott's evocative descriptions provide a lens through which one might view Indian fables about magical lakes and other bodies of water as observations of Fata Morgana. The ancient text *Adbhutasagara*, so P. V. Kane records in his mighty compendium of ancient and medieval Indian sources, catalogues atmospheric phenemona, including Fata Morgana.[13] The exceptional colonial ethnographer William Crooke records the legends that attach to the Taroba lake in central India, tales that seem to describe complex mirages. One narrates how a marriage party passing by Taroba was dismayed to find no water. A curious elderly figure urged the groom and bride to join him in digging for water and they were soon able to slake their thirst. However, the water rose and rose, drowning the couple. Then 'fairy hands soon constructed a temple in the depths, where the spirits of the drowned are supposed to dwell.' Subsequently, a palm tree grew on the side of the lake. This was visible only during the day, 'sinking into the earth at twilight'. Then, one day, a pilgrim sat on the tree and 'was borne into the skies, where the flames of the sun consumed him'.[14] When the waters sank low, 'golden pinnacles of a fairy temple were seen glittering in the depths'.[15] Crooke sees in elements of these stories evidence of a 'genuine solar myth', but they resonate in striking ways with the mythology that surrounds Morgan le Fay, King Arthur's sister who dwelt in an underwater palace.[16]

Fata Morgana's ground zero is the Straits of Messina, the water that separates the eastern tip of Sicily from Calabria. Home to whirlpools thought to be the origin of the Scylla and Charybdis myths, the Straits have long been the site for the appearance of a complex superior mirage, often with elaborate architectural features. Although known

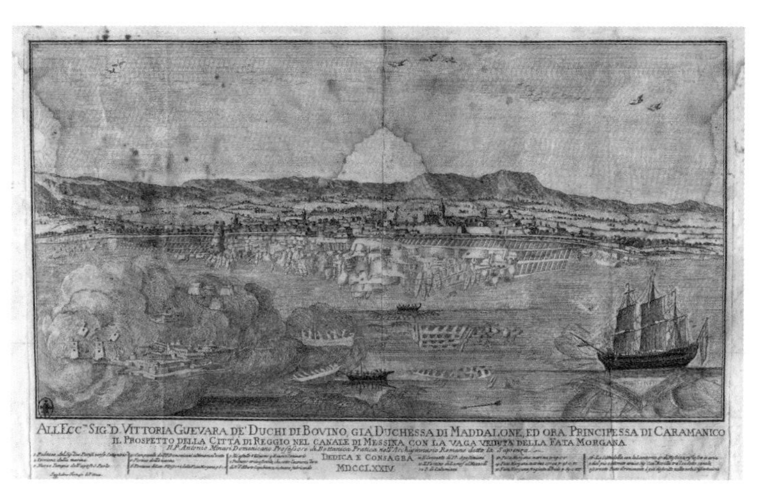


4 Guglielmo Fortuyn, after a description by Antonio Minasi, *Prospect of the City of Reggio in the Straits of Messina with a View of the Fata Morgan*, 1773, copper engraving with woodcut initials.

to the ancients, the earliest documented account is given by Fazello in 1558, but the classic early description of this (dating from 1773), is by the Dominican friar Antonio Minasi, and would be anthologized in numerous later accounts.[17] It describes

> various multiplied objects such as a numberless series of pilasters, arches and castles well delineated, regular columns, lofty towers, superb palaces with balconies and windows, extended alleys of trees, delightful plains with herds and flocks, armies of men on foot and horseback, and many other strange figures, all in their natural colors and proper action, and passing rapidly in succession along the surface of the sea, during the whole short period of time that the above-mentioned causes remain.[18] (illus. 4)

This astonishing vision appears 'in the water' as a marine mirage and it is this which the most famous early depiction records (although as the meteorologist C. Fitzhugh Talman notes, this reverses the spectator's position, presenting a kind of ideal diagram: the mirage was always seen *from* the city of Reggio, which appears here as a backdrop).[19] This Ur-Fata Morgana also manifested as an aerial (actually 'superior') mirage when 'the atmosphere be highly impregnated with vapor and exhalations not dispersed by the wind nor rarefied by the sun', producing effects similar to that shown in a wonderfully imagined chromolithograph in the German compendium *Die Wunder der Natur* (1913), in which we see a Sicilian sailor gazing at a magical towering apparition, a fluttering airborne concatenation whose original elements are all clearly discernible in the shoreline below (illus. 5).[20]

Minasi likened the mirage to a 'Catoptric Theatre'. Catoptric theatres had been popularized by the Jesuit Athanasius Kircher's

5 After Henry Charles Seppings Wright, 'A Fata Morgana (Mirage) in the Straits of Messina', chromolithograph in *Die Wunder der Natur* (1913). The image perfectly illustrates Henry Swinburne's description in 1803 of a 'marine looking-glass; which, by its tremulous motion, is . . . cut into facets.'

*Ars Magna Lucis et Umbrae* (The Great Art of Light and Shadow), in which he had illustrated a folding box lined with mirrors in such a way that they replicated reflections to infinity.[21] Significantly, Kircher had himself travelled to Reggio in the (to be thwarted) hope of seeing the mirage, and would publish one of the earliest accounts of the Fata Morgana in 1646. Kircher cites a letter from Father Ignazio Angelucci, who had seen the mirage in 1634:

> The sea that washes the Sicilian shore swelled up, and became for ten miles in length, like a chain of dark mountains; while the waters near our Calabrian coast grew quite smooth, and in an instant appeared as one clear polished mirror, reclining against the aforesaid ridge. On this glass was depicted in *chiaro scuro*, a string of several thousands of pilasters, all equal in altitude, distance and degree of light and shade. In a moment they lost half their height and bent into arcades, like Roman aqueducts.[22]

Kircher was of the belief that the spectacle was created by tiny airborne crystalline fragments of rock and sand, so much so that he collected samples to take to Rome, where he heated them in a vessel made into the shape of the Straits of Messina before passing a light through the haze in a (surprisingly successful) effort to generate a mirage.[23] This endeavour he later repeated in front of sundry ecclesiastics, with the addition of projected 'moving warriors and devils upon a screen'.[24] Minasi, by contrast, believed that the strong countercurrents in the Straits led to a longitudinal difference in the level of the water 'so that it behaved like a mirror lying, not horizontal but tilted at a slight angle . . . and reflecting objects along the Calabrian shore'.[25]

TWO

# A World History of Mirages: The Thirst of the Gazelle

This journey, from nineteenth-century colonial India to medieval Sicilian reworkings of Arthurian legend, may well be worthy of inclusion in Wallace's misleading cyclopaedia. Mirages truly occupy a catoptric box in which every account and experience of one finds its mirror echo in some other place or time. Like Wallace's cyclopaedia, one entry seems always to refer to another, which in turn leads to another. Sometimes, this produces a slippage, or *différance*, a fugitive will-o'-the-wisp worthy of the catoptric box. At other times we encounter a 'citationality' characteristic of scientific description through which anomalous phenomena are placed within a matrix of other similar observations in order to establish similarities and differences. One senses the slipperiness of mirage provoking attempts to capture it through linkage and connections within a grid of reverberating echoes.

Tod's movement from India to the Old Testament to Quintus Curtius describes a trajectory mirrored in many other early accounts of this evanescent and only-just-named phenomenon. These early musings are often driven by a feeling of wonder and a sense of crisis about forms of knowledge that depend upon visual certainty. An understanding of 'natural magic', and a sense of the universal medium of 'atmosphere' susceptible to global science, serves potentially to position mirages within what we might think of as a 'de-territorializing' frame. This potential was most easily realized with respect to polar mirages, for they were much more difficult to situate within the cultural understandings of local populations and much easier to liberate in the global space of science. Many de-territorializing accounts invoking that global space, for sure, dip in and out of local folklore, searching for insight, but this is knowledge channelled into a globally connective strategy. In contrast, we will also encounter other kinds of accounts that we might think of as 'territorializing'. These tie the appearance of mirages to specific locations and cultural practices. Sometimes, as in India or China and Japan, the intent is to place the mirage within a deep civilizational tradition and to demonstrate how myth and poetry have responded to the ephemeral and mysterious. We might think of these as strategies of enchantment. Elsewhere, most notably in deserts (what Kipling memorably condemned as the 'sand-bordered hell' of West Asia and North Africa),[1] mirages signify something more ambivalent. Islamic West Asia and North Africa emerge as the natural home of mirages as much because of European anxieties about

6 A Fata Morgana in a desert in Africa, an early Italian encyclopaedia illustration based on a watercolour by Friedrich Perlberg.

7 B. K. Mitra, 'Mrigtrishna', or the 'thirst of antelopes', from the journal *Kalyan* (*c.* 1950). The deer meet their end in the sand dunes under a blazing sun above three kinds of human desire (*kammohit*, the allure of lust; *lobhmohit*, the allure of avarice; *madmohit*, the allure of intoxication).

Islam as because of the particular heat gradients to be found in desert environments (illus. 6). Here mirages seem to mediate fundamental contests of vision, and questions of transparency and occlusion, that European travellers systematically saw as intensified by Islam. This is further complicated by a long European preoccupation with Ottoman 'dreaming' and an even longer Islamic tradition that opposes the truth of dreams to the falsity of everyday sense perception.[2] In the cultural contact zones where mirages appeared, there was frequently more than one narrative about the significance and nature of mirage.

Often local narratives are folded into the scientific repertoire. We can recover an Indian vision of the phenomenon via another more global text with which James Abbott might well have been familiar. This was Alexander von Humboldt's *Views of Nature* (1834), whose brief reference to mirage illusions was significantly expanded in his 'illustrations', the empirical appendix that makes up the bulk of the book. Humboldt's main text remarks on the 'thirsting wanderer . . . deluded by the phantom of a moving, undulating, watery surface, created by the deceptive play of the reflected rays of light'.[3] Although Humboldt was empirically focused on his observations in the Americas, his editor also included a footnote referring to a plate in Robert Melville Grindlay's *Scenery of the Western Side of India* (plate 18, which 'represented well [a] narrow stratum' between the ground and palm trees 'hover[ing] aloft' due to the heat and density of air, an example of a superior mirage, and also the Indian motif of the 'thirst of gazelles', an example of an inferior mirage). Humboldt's supplementary commentary noted that in Sanskrit 'the mirage is called "the thirst of the gazelle"' before also mentioning the French Expedition to Egypt (where 'this optical illusion often nearly drove the faint and parched soldiers to distraction', and to which we owe the introduction of the term 'mirage'), and Diodurus Siculus' description of the 'strange illusive appearance' of an African Fata Morgana.[4]

For Humboldt, these dispersed examples served to underline 'the sublime phenomenon of creation' (the subtitle of his book), a phenomenon that was global and interrelated. Humboldt pursues a cosmic 'worlding' of phenomena that folds natural sublimity and human

8 *The Enchanted Deer*, lithograph published by Calcutta Art Studio, *c.* 1880. Sita points to the illusory deer, Ram sits with a bow to her right and Lakshman stands to her left.

responses to it into his unifying biogeography.[5] But what for Humboldt were centripetal forces can also, in other frames, assume centrifugal trajectories. The image of deer or gazelles searching for illusory water in the desert is occasionally used to illustrate the Sanskritic concept of *mrigtrishna* (a concept recognized today in the village where those Miraj tobacco packets are sold: in that village, Mrigrtrishna is thought to be an oasis in Rajasthan).[6] However, as a mid-twentieth-century illustration from the pious (indeed reactionary) journal *Kalyan* clearly demonstrates, the 'thirst of gazelles' is a metaphor for human longing for what can never be found (illus. 7).[7] In the *Kalyan* image, the deer meet their end in the sand dunes under a blazing sun above an analogical register of human desire (three kinds of *mohit* denoting 'allure'). In the middle section of the image (by Gita Press resident artist B. K. Mitra) we see an adulterous couple, whose passions need to be cooled by an electric fan, under the heading *kammohit* (the allure of lust); a greedy pipe-smoking merchant surrounded by sacks of cash under the sign of *lobhmohit* (the allure of avarice); and two men, smoking and drinking behind closed curtains, secretly, under the sign of *madmohit* (the allure of intoxication).

In India the connection between deer, desire and illusion is deep. Valmiki's *Ramayan* pivots one of its most dramatic episodes around the disguise of the *rakshas* or demon Maricha (whose name conveys the sense of 'mirage') in the form of a beautiful deer for whose skin the exiled Sita yearns. Ralph Griffith's English translation notes Maricha's 'magic power' and 'rapid change', before describing how:

> He wears, well trained in magic guile,
> The figure of a deer a while,
> Bright as the very sun or place
> Where dwell the gay Gandharva race.[8]

The *Ramayan*'s enfolding of disguise, illusion and desire (also signalled by the reference to the Gandharvas, demons whom we will shortly encounter as the creators of Fata Morganas) directs us to the moral pedagogy that lies at the heart of Indian thought on mirages.

The *mrigtrishna* would achieve a certain popular literary celebrity through the efforts of the remarkable early twentieth-century author and professor of history at the Deccan College in Poona, F. W. Bain (1863–1940), who published ten books claiming to be translations from Sanskrit. His first book, *A Digit of the Moon*, published in 1899, was catalogued in the Oriental Department of the British Library.

Part two of *Bubbles of the Foam* is titled 'The Thirst of an Antelope' and the frontispiece depicts a solitary deer in a desert gazing towards a waterless sea (illus. 9). Bain proclaims that the desert is one of the four fundamental locations of 'old Hindoo literature' and suggests that 'old Hindoo people' discovered when they tried to cross deserts that 'above the plain prosaic danger, this Waste of Sand laid . . . goblin snares for the unwary traveller's destruction, in the form of its Mirage.'[9] The 'goblin snares' were a product of their ignorance of 'optical phenomena' and involved 'phantom trees and water, these mocking semblances of cities that vanished as you reached them'.[10] These phenomena struck their imaginations and they gave them poetic names such as 'deer-water' and the 'thirst of the antelope'.

Bain then perceptively interprets these mirages as images made perfectly in the shape of 'moral meanings' and 'philosophical ideas'.[11] Chief among these is *maya* or illusion, which Bain reads as 'metaphysical Delusion, in which all souls are wrapped, which leads them to impute Reality to the Phantasms'.[12] The illusion/delusion encourages its victims to keep hunting for what cannot be found:

9 'The Thirst of an Antelope', frontispiece from F. W. Bain, *Bubbles of the Foam* (1912). Bain's hugely popular spoof Sanskrit translations were strikingly respectful exercises in popular Orientalism.

> Mirage! mirage! . . . The world is unreal, a delusion and a snare; sense is deception, happiness a dream; nothing has true being, is absolute, but virtue, the sole reality; . . . We move, like marionettes, pulled by the strings of our forgotten antenatal deeds, in a magic cage, or Net, of false and hypocritical momentary seemings: and bitter disappointment is the inevitable doom of every soul . . .[13]

Thirsty gazelles search for water conjured by inferior mirages. But superior mirages have an equal place in the Indian tradition's preoccupation with illusion and morality. What we would now rationalize as Fata Morganas have a very deep history in Indian literature, where they appear as the 'city of Gandharvas', which Wendy Doniger O'Flaherty translates as 'magic city in the sky'.[14] The Gandharvas (mentioned, as we have seen, in the *Ramayan*) are demons 'who are masters of illusion and deception', remarkable magicians capable of conjuring vast aerial conurbations (illus. 10). Cities signify the 'false pride' of men who believe they are capable of building enduring structures and also feed into a corpus of texts that deal with the 'nightmare of solipsism'.[15] The *Yogavasistha* describes a city in the sky, as 'insubstantial as a cloud' and inhabited by a 'fool' who 'lived all by himself in an empty place, like a mirage in a desert'. Inevitably the city fades, and so he 'built another, and another, and another, and all of them dispersed into the air'.[16] Modern physics may emphasize the different particular atmospheric conditions that result in inferior as opposed to superior mirages, but the Indian tradition grasps both as equally powerful moral metaphors. The city in the sky delivers the same moral message as the deer in the desert: desire magnifies the illusory. But the Indian tradition, while seeking liberation from illusion, understands how persuasive illusion is. We might say that it respects illusion's illusory power. There is no

10 'Mirage', from G. Hartwig's *The Aerial World* (1874). This astonishing vision of an aerial coastal settlement makes tangible the Indian vision of the city of Gandharvas or 'magic cities in the sky'.

simple Platonic condemnation of 'shadows' but rather an appreciation of the phenomenal power of multiple universes. The *Lankavatarasutra* sagely notes that a city

> is neither a city nor not a city . . . It is just like some man who is asleep in bed and dreams of a city with its women, men, elephants, horses, and so forth . . . and who wakes up just as he enters the inner apartments of the palace.[17]

The Gandharvas' key building blocks are clouds,[18] and Doniger O'Flaherty draws attention to Shakespeare's exploration of clouds as shape-shifting deceivers, not only in *Antony and Cleopatra* (IV.14) but in an exchange between Hamlet and Polonius:

> *Hamlet*. Do you see yonder cloud that's almost in shape of a camel?
> *Polonius*. By the mass, and 'tis like a camel, indeed.
> *Hamlet*. Methinks it is like a weasel.
> *Polonius*. It is backed like a weasel.
> *Hamlet*. Or like a whale?
> *Polonius*. Very like a whale.[19]

We will encounter the relation between weasels, whales and visual interpretation again below, in a way that returns us to India, in relation to the Warren Hastings trial of 1788.

THREE

# 'Fallacious Evidence of the Senses'

One of the greatest insights of the twentieth-century German critic Walter Benjamin concerned what he termed 'the optical unconscious', to which photography gave us access. His idea was that there was a set of visual truths that ordinary vision was unable to grasp. Film (both still and moving), by contrast, was able to juxtapose, magnify and slow down an otherwise intractable reality. Photography disrupted an otherwise unchallengeable order and offered new tools of resistance.[1]

Mirages, by contrast, were part of an optical fallaciousness. Optically 'real', but not 'true', they disordered experience, tricked their beholders and provided proof that sense experience was not to be trusted. An anonymous review of David Brewster's *Letters on Natural Magic* in the Catholic periodical *The Dublin Review*, probably by the liturgist Daniel Rock, marshalled evidence of the persuasive nature of mirages (a delusive persuasiveness that even men of science could not deny) as proof that the senses 'are utterly incompetent and inadmissible as faithful guides in any investigation on the mysteries of religion and objects of divine faith'.[2] Mirages, together with other mysteries (such as portraits whose eyes follow the viewer round a room), were conclusive evidence of the 'fallacious evidence of the senses' and the 'discordance between our perceptions and their causes'.[3]

Rock's target was Brewster and the tradition he embodied. As visual puzzles and zones of uncertainty, mirages were important to a philosophical tradition for which the eye was granted a privileged place as the arbiter of knowledge.[4] The eye was central to Brewster's understanding of the world, the optic nerve being the means by which 'the mind peruses the hand-writing of nature'.[5] A scientist in the Scottish common-sense philosophical tradition, and inventor of the kaleidoscope, Brewster was interested in 'illusions as experiments in the process of making knowledge'.[6] His *Letters on Natural Magic* of 1832 sought to demonstrate that 'susceptibility to deception was built into the mechanism of the eye' and that the scrutiny of this fallibility would deepen our understanding of these mechanisms.[7]

Rock opens his contrasting lesson in the misplaced faith that we invest in our senses (when instead, he implies, we should place more faith in 'the mysteries of belief' and the 'inexplicable truths of revelation'[8]) with a very detailed and entertaining account of the imaginary island of St Brandan, which has persistently 'haunted the imagination of the inhabitants of the Canaries'. Canary Islanders imagined that they saw a mountainous island, ninety leagues in length, lying to the west. It appears in Martin Behaim's globe of 1492 (the Erdapfel, now in the German National Museum in Nuremberg). In an *Encyclopédie Larousse* abstracted illustration of the globe of 1898, St Brandan appears as a huge land mass hovering just above the equator (illus. 11). Most maps at the time of Columbus, Rock notes, placed St Brandan 'about two hundred leagues west of the Canaries', but it had a much deeper history, being known to the ancients and referred to by Ptolemy as Aprositus ('inaccessible'). For Rock, it is important that this is no fleeting deceit but a systematic deception of the senses of the mass of the populace over the centuries, suggesting a profound inability by (ordinary) humans to determine, on their own, what is true or false.

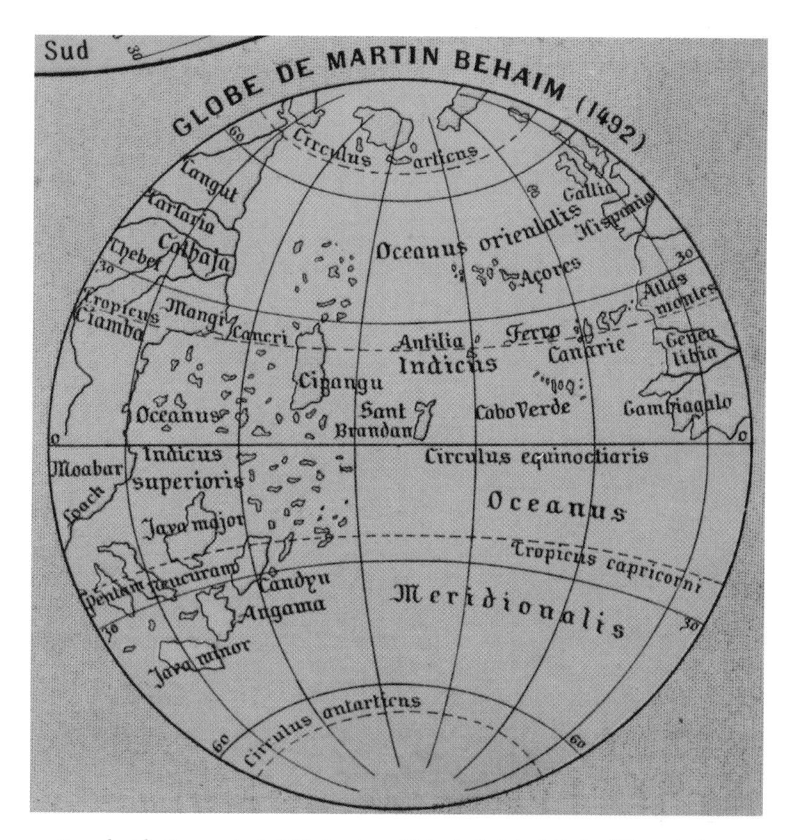


11 *Encyclopédie Larousse*'s 1898 illustration of Martin Behaim's *Erdapfel* (1492), showing the mirage island of St Brandan.

Hence Rock is keen to stress that the island was 'repeatedly seen', by many observers at the same time, and always in the same place 'and in the same form'. Fernando de Troya and Fernando Álvarez led an expedition in pursuit of it in 1526 and of course found nothing. By this stage, however, such was the power of the phantasm's 'secret enchantment for all who beheld it, that the public preferred doubting the good conduct of the explorers rather than their own senses'.[9] Subsequently one Alonzo de Espinosa, Governor of Ferro in the Canaries, filed a report in which more than a hundred witnesses testified to their sighting of the island.

This, together with other accumulated proofs, prompted another expedition in 1570, commanded by Fernando de Villalobos, which departed from Palma. It, too, failed to locate the mirage. Rock almost purrs with delight: 'St Borondon seemed disposed only to tantalize the world with distant and serene glimpses of [an] ideal paradise . . . but to hide it completely from the view of all who diligently sought it.' So of course a further expedition was launched in search of the Canary Islanders' 'favourite chimera.'[10] In 1605 Gaspar Pérez de Acosta led a voyage that was also unsuccessful. A final, fourth expedition led by Don Gaspar Domínguez in 1721 was provoked by the 'lemons . . . and green branches of trees' that washed up on the shores of Gomara and Ferro and that were assumed to have originated in the 'enchanted groves of San Borondon'.[11]

Because the public did not listen to wise men of science (just as they did not listen to those who understood recondite matters of religion), St Brandan refused to disappear. It appeared as one of the Canary Islands in a French map of 1704, and in another in 1755, where it was placed five degrees west of Ferro. It became indestructible: 'It was in vain that repeated voyages and investigations proved its non-existence,' and the public, 'to defend their favourite chimera', sought refuge in the supernatural. It became muddled with the Seven Cities, where seven bishops had taken refuge from the Moors, with the abode of the Portuguese King Sebastian, or that of the Spanish Roderick, also fleeing the Moors. Some thought it a 'terrestrial paradise', a place 'made at times apparent to the eyes, but invisible to the search of mortals'.[12]

Rock's Catholic conclusion is that 'common sense' (of which David Brewster was a champion) cannot be relied upon. The history of the persistence of St Brandan demonstrated that. Just as, in matters of science, we must rely on 'a few gifted and educated individuals' and accept 'upon trust' through 'human faith' what we cannot understand (and which the evidence of our own sense may well contradict), so also in religion the 'unlettered' must once again discover the necessity of accepting matters that may appear 'apparently incompatible with anything like possibility'.[13]

Whereas Benjamin's optical unconscious was a revolutionary tool of disenchantment, for Rock the 'optical uncertainty' embodied in the mirage taught a very different kind of reactionary lesson. For Benjamin, the optical unconscious promised to destroy aura and the 'cultic'. For Rock, mirage's optical confusion demonstrated the limits of democratic reasoning and the necessity of the restoration of hierarchical interpretation in matters of both science and religion. Rock clearly had a theological axe to grind, and mirages were actors in his drama staged on behalf of the Catholic hierarchy.

'Science' narrated a very different kind of story in which experiment and experience overcame faulty belief. It is therefore no surprise that mirage often appears as a faulty sign to be corrected by proper geographical or metrological observation. An early articulation of this position occurs in Hibbert's entertaining 1826 work on apparitions (which he elsewhere referred to as 'spectral impressions') that deployed science to explain religious delusion. 'Ghost stories', he declared, are often

> referable to optical mistakes of the nature of external objects. The phenomena [such as] the Giant of the Bro[c]ken are known to everyone. To the same class of *pseudo-apparitions* belong the Fata Morgana, and the Mirage or Water of the Desert.[14]

Sometimes this scientific triumphalism twisted the evidence in pursuit of a good story. A front-page story in *Scientific American* in December 1856, for instance, gleefully reported one such narrative of redemption. It involved dubious secondary sources (including

*The Westminster Review*'s partial account of C. J. Andersson's *Lake Nagami*, which had appeared earlier that year).[15] The claim was that Andersson, who was travelling with Francis Galton, had 'put an end to a lie', communicated by Arab travellers to missionaries that there was a 'vast fresh water lake' in the interior of Africa. This was described as 'of enormous dimension', and as 'nothing less than a great inland sea'. *Scientific American* also reported that 'frequenters' of the Geographical Society's meeting in London had observed the appearance on the Society's map (where previously there was desert) of 'a blue spot, about the size of the Caspian, in the shape of a hideous inflated leech'. Andersson's text reveals this to be Lake Omanbonde, which had evaporated due to a profound drought. For *Scientific American*, the case was shut: 'The lake turns out to be mirage – a mythus with the smallest conceivable nucleus of fact.' It was not a great blue leech but something more like the size of Lough Corrib, in Galway.[16]

Almost a century and a half later, science still appears in the ascendancy, deploying its knowledge of refraction to explain medieval mysteries. The thirteenth-century *King's Mirror* describes three mysterious features of the Greenland and Iceland Seas – the *hafgerdingar*, the merman and the *hafgufa* or kraken. The first of these had been subsequently explained as a 'sea fence', that is 'a visual effect created by anomalous atmospheric refraction' and this prompted the Canadian engineers Lehn and Schroeder to revisit the evidence concerning mermen (essentially male equivalents of mermaids, human male from the waist up and – possibly – fish from the waist down). From the seventeenth century, mermen (and mermaids) were frequently rationalized as dugongs or manatees. With the aid of several graphs of temperature profiles and a lot of equations, Lehn and Schroeder demonstrate how the heads of killer whales or walruses are very likely the real objects which atmospheric refraction presented as merman mirages.[17]

## FOUR

# 'Mocking Our Distress'

The 'scientific' response, in which mirages were conjured as mysteries so they could be dissolved by reason, was often thwarted. In part this was due to insufficient and inaccurate information. As late as 1914, a correspondent in *Scientific American* bemoaned the poor documentation of arctic mirages despite the polar regions being home to unequalled 'remarkable forms of mirage'. The correspondent approvingly notes the explorer William Scoresby's description of a Fata Morgana off the coast of Greenland at the beginning of the nineteenth century, which compare it to 'an extensive ancient city, abounding with the ruins of castles, obelisks, churches and monuments'. 'The whole exhibition', Scoresby continued, was 'a grand and interesting phantasmagoria'. The reports of other travellers were feeble by comparison. 'Awkward circumlocutory descriptions of the phenomena are substituted for their names,' these being akin, the correspondent continued, to a traveller returned from the Sahara who reports seeing a 'large, brown quadruped with a hump on its back'. Scott's expedition took with it *The Antarctic Manual*, an 'elaborate book of instructions', but it contained no mention of optical phenomena. *Scientific American*'s correspondent concludes with praise of Alfred Wegener's photographs as published in the reports of the Danish Greenland expedition of 1906–8 and the evidence they produced of the analytic and descriptive promise of the camera.[1]

The power of mirage lay in its extraordinary presence and its insistence on its own account of reality. Time and time again beholders of the mirage describe how they know what they behold is an optical effect and not a true version of the world, and how the visual data continually threaten this scepticism. Off the coast of Brazil, during his circumnavigation between 1764 and 1766, Vice-admiral John Byron describes an island 'plain as ever I saw . . . in my life'. Changing his ship's direction produced no change in the aspect of the island. It continued to look 'very Blue', with some of the crew claiming they saw the sea breaking upon its 'Sandy Beaches'. And then it suddenly disappeared, to Byron's 'great astonishment'. 'Tho' I have been at Sea now 27 Years & never saw such a Deception before . . . I dare say there is not a Man on board but would have freely made Oath of the certainty of its being Land,' Byron concludes.[2] Fifty years later the Irish abolitionist R. R. Madden, encountering a mirage ten miles from Suez, addresses this conundrum very directly, although he seems confused about the role of the imagination:

> If I were to speak of the nature of *Mirage* from my own sensations, I should say, it was more of a mental hallucination rather than a deception of the sight; for, although I was aware of the existence of the Mirage, I could not prevail on myself to believe that the images which were painted on my retina were only reflected, like those in dream, from the imagination, and yet it was so.[3]

A further response, one that runs through numerous travel accounts, attributes a kind of malevolence to the deception by mirage. Clearly this is in part a genre convention of saleable travel narratives (an aspect that Michel de Certeau termed 'storms and monsters'), and mirages very quickly establish themselves as part of the inevitable confrontational dramaturgy of the desert.

The soldier and future Governor of Hong Kong Henry Pottinger's 1816 account of Baluchistan and Sind (in what is now Pakistan) fuses a knowledge of optics with a vivid topography, the angel dust of local (linguistic) knowledge, and the theatrical depiction of the thirsty traveller fooled by refraction. The *suhrab* or mirage is 'caused by the refraction of the atmosphere from extreme heat' and derived (as he documents in a footnote) 'From Suhr, the desert, and Ab, water'. Floating all around the travellers we find this *suhrab* 'mocking our distress by its delusive representation of what we so eagerly thirsted for'.[4] This theme was amplified a few years later in the *Penny Magazine*'s consideration of the cruel fact that 'it is precisely in those districts where the traveller is exposed to the most intense agonies of thirst, that his wants are mocked by the illusion' of mirage. This contrast and the paradox enthused the writer:

> Every eye is dim; every tongue swollen, parched, and rent, cleaves to the roof of the mouth; and the Arabs begin to talk of killing the camels for the sake of the water contained in their stomachs. In such circumstances it is easy to imagine the delight with which, in the heat of the day, the traveller perceives before him one or more lakes, reflecting on their clear surface . . . palm-trees . . .[5]

By 1903, the travel writer E. A. Reynolds-Ball was urging travellers (in a strangely military language) to the Near East to use their rifles to judge what was water and what was not: 'Mirage: how to detect. If you sight the supposed lake with a rifle, you will detect movement, which will betray its unreality.'[6] This book by Reynolds-Ball was very closely studied by T. E. Lawrence in preparation for his own adventures.

Commonly the 'fraud' of mirage is sufficiently compelling that the traveller is seduced. J. E. Richter, in a report titled 'Effects of Thirst', recalled his near-death experience in a 'dreadfully hot' desert of Central Australia in 1864. The pint of water he took to be shared with his horse is soon finished and his tongue sticks to his mouth. Time drags slowly over these 'weary cheerless plains' until a 'flaring mirage' springs up on the horizon, 'inciting' in him 'the hope that it was a lake of water . . . though I knew quite well that no lake existed in that direction'.[7] An equally memorable version of this will to believe is given by F. L. James in his account of travels in Muslim East Africa from the 1880s:

> [en route to Hodayu] The water will not hold out two days. Where are the wells? No one has the least idea. During eight days we have been marching through tropical heat, and have divided one pint of water between five of us for washing purposes. Into this we dipped our faces in turn, and then dipped our hands together, each scooping into his palm what he could of this rich solution of himself and four others [sic]. Ahead we see a misty lake fringed by green trees, but the trees are upside down, so it is only a wretched mirage, with a very feeble deceptive power. As we approach this fraud, we become aware of a thin belt of mimosa, which is distinctly green and real, and we almost dare hope water may be at hand; but the hope is rapidly put aside for the country beyond is, if possible, more dead and desolate than ever.[8]

This quality of the mirage prefigures questions that emerged later in the nineteenth century about magic and cinema. Many concerned with magic stage performances debated 'the extent to which the audience should be allowed complicity with the workings of the trick'. Similarly, in film the question focused on whether the audience should be let into the secret: 'should the projector be out of the audience's view, deepening cinema's illusionism, or should it be on display, making the technology itself central to the show?'[9] The issue of illusion was also central to what the early film historian Tom Gunning considers the founding myth of cinema, the claim that the audience in the Grand Café ran fleeing for their lives when confronted with the Lumière brothers' short *L'Arrivée d'un train en gare de La Ciotat* in 1895 or 1896. Such myths can be seen as a way of keeping alive a kind of savage mimesis – as modern spectators, we deny such mimetic 'mistakes', but we nevertheless demand the presence of non-modern spectators who can testify to, and authorize, the extraordinary power of the image.[10]

Murray Leeder, developing Simon During's observation that modern culture, like secular magic, is built upon 'illusions *understood as illusions*', draws the important conclusion that magic creates a '(half-way) knowing spectator', a spectator 'who is allowed a certain kind of complicity in the magicians' technique'.[11] This perfectly describes the ambivalent position of mirage spectators, who find themselves seduced by what they know to be illusory, as so many accounts reiterate. They share a complicity in the mechanics of mirage – they understand that it is the result of refraction contingent on the differential density of air. But the act of beholding involves an erasure of this distantiating knowledge. Often the adventurer or explorer narrates to themself their scepticism at the very moment that they concede its undoing. They can rationalize the optics of mirage but to the extent that they are seduced by its wonder, they succumb to magic.

Placing mirages, and the self-conscious narratives about their status, in this context allows us to grasp the complexity of the moral and political issues that they signify. It is not simply that in their occlusion and visual deception they point to what is deemed by the European adventurer to be negative (the desert, thirst, despotic politics, Islam), for they also point to a desirable illusionism, a state of not knowing, of sublime obscurity, that highlights the seduction of illusion. Mirages point to a *cogito* that ultimately has no principle of certainty. The numerous narratives that declare the optics and physics of mirages before conceding seduction pose a thrilling challenge to Cartesian certainty. These narratives suggest that there is no knowledge which can withstand dream or illusion.

## FIVE

# Cold and Hot: The Geography of Mirage

Mirages have been observed in any environment where there are sufficient temperature gradients, or temperature inversions, to generate the necessary refraction. We have already recorded the foundational role of apparitions in the Straits of Messina in the history of mirage. David Brewster's *Letters on Natural Magic*, like Humboldt, makes clear the geographical ubiquity of mirages (Brewster lists appearances at Cumberland, Dover, Hastings and elsewhere). Further, there are enough descriptions of arctic Fata Morganas to fill a whole book whose subject might be complex cold-weather superior mirages (illus. 12 and 13).

If this were such a book, we would find a place for William Scoresby's experiences on 16 July 1814 as he sailed past Charles Island, Spitzbergen (Prins Karls Forland), high in the Arctic Ocean. He memorably described what appeared to be a mountain, a surprising one, for he had never seen it before. More astonishing, however, was a 'prodigious and perfect arch'. The mirage (the result of 'looming') then changed: 'the mountains along the whole coast, assumed the most fantastic forms; the appearance of castles with lofty spires, towers and battlements, would in a few minutes, be converted into a vast arch or romantic bridge.' Scoresby is clearly astonished – 'these varied and sometimes beautiful metamorphoses, naturally suggested the reality of

12 Postcard depicting 'Fata Morgana at Sea', Grimme & Hempel, Leipzig, *c.* 1900.

13 *Delibab a horlobagyon, Fata*, Hungarian postcard, *c.* 1910. In Little Cumania, M. Malte-Brun's *Universal Geography* of 1828 noted: 'the mirage is often seen in the hot days of summer; it is the *deli baba* or fairy of the south, it tantalizes the shepherd and his thirsty flock with the sight of azure lakes crowned with forests, palaces and ruins.'

fairy descriptions' – but he insists upon the objectivity of these visions. They were 'uncommon phantasms' but Scoresby stresses their plausibility: even when examined with 'a powerful telescope', the mirages 'seemed to possess every possible stability'.[1] Scoresby's protestations notwithstanding, his elaborate description, like Tod's and Abbott's, draws our attention to the manner in which the beholder's imagination fills in the detail provided by the mirage template.

The 'looming' mirages for which Scoresby became best known involved ships of the *Flying Dutchman* variety. Under certain circumstances, he notes, 'all objects seen on the horizon seem to be lifted [and] extended in height above their natural dimensions.' Often appearing to be connected to the horizon by 'fibrous' or 'columnar' extensions,[2] these ships, elevated and also often inverted, became a favourite subject for engravers of nineteenth-century scientific encyclopedias (illus. 14).

Scoresby was always in control of his mirages: he could unveil them as optical wonders of nature without any danger of undermining his authority. The Royal Navy Commander John Ross's experience during the British Admirality's Arctic Expedition was less happy. William Edward Parry, Ross's second-in-command, wrote of 'the exhalations arising from the earth [that produced] that appearance of waving tremulous motion in distant objects, which the French call *mirage*'.[3] This scepticism was put to use when confronted during the search for the Northwest Passage with a huge landmass that Ross decided to name Croker Bay (in honour of J. W. Croker, the Secretary to the Admiralty, thus ensuring his subsequent humiliation) before concluding there was no passage.[4] Parry, phrasing the matter rather delicately, noted: 'I have called this large opening a bay, though the quickness with which we sailed past it did not allow us to determine the absolute continuity of land round the bottom of it,' and he concluded that it was 'by no means improbable, that a passage may here be one day found from Sir James

Lancaster's Sound into the Northern Sea'.[5] Parry would do exactly this the following year, sailing through the Fata Morgana. Having nailed the name of the Secretary of the Admiralty to an illusion, Ross failed to secure any further command. One subsequent commentator ascribed Ross's error to 'a kind of second sight, peculiar to his countrymen, who sometimes see things, which have no existence in reality'.[6]

Other parts of North America also provided fecund mirage locations. A correspondent in *Scientific American* in 1868 recorded how a friend 'was crossing the western plains and saw distinctly a broad stream, fringed with trees, and having dwellings on its banks, so plainly described and fairly presented that he urged his horse on to reach what, to him, was a paradise, but found only bare sand'.[7] This description is almost the literary mirror to Alfred Jacob Miller's watercolour record of the overland trail through western Wyoming, *Mirage on the Prairie* of 1858. Emily Wilson notes the 'icy, jewel-toned palette and hazy

14 'Mirage at the North Pole (Expedition of the *Germania*, 1869)', woodcut from Flammarion's *L'Atmosphère* (1873). The *Germania* was part of the Second German North Polar Expedition.

atmosphere', which she suggests is indebted to J.M.W. Turner's *Pyramids at Gizeh*. These 'magnify the feeling of distance and create a sense of spatial distortion, not unlike the experience of marching toward a mirage' (illus. 15).[8]

The far west veritably heaves with mirages in a wonderful catalogue of sightings and lore relating to ghost riders published in *Bentley's Miscellany* in 1837. This article, probably authored by Charles Fenno Hoffman, is in part extracted from an earlier piece in *American Monthly Magazine* titled 'Nights in an Indian Lodge', which combines proto-ethnographic passages with lurid frontier narrative.[9] The 'illusive [inferior] mirage of the desert cheats the parched traveller with its refreshing promise', but it is superior mirages that give rise to 'monstrous shapes and unnatural forms', which 'reflected on the heated and tremulous vapour, are magnified and distorted to the eye of the appalled and awe-stricken traveller'. Fire shoots along the 'baked and cracking earth' and wild horses appear on the horizon 'goaded on by gigantic and unearthly riders, whose paths are enveloped in wreaths of flame'.[10] The 'mocking mirage' persuades Native Americans and Creole hunters of the identity of the ghost riders whose names are whispered with awe. The Omaha warrior 'invariably places his hand upon his Metawaüann, or repository of his personal manitto, when he speaks of these fearful apparitions'.[11] The ghost riders are two gigantic figures. Those who claim to have seen them close-up (though they 'mock the scrutiny of human eyes') report that they are a white man ('emaciated and ghastly, and writhed with the most fearful contortions by an expression of shrinking horror'), and a Native American female ('collapsed and corpse-like').

The identity of these two riders and the figure that forces them ever onwards unfolds in the story 'Ghost-Riders' through a narrative of 'Eden-like happiness blasted by unholy passion'. Ta-in-ga-ro

15 Alfred Jacob Miller, *Mirage on the Prairie*, 1858. Miller's watercolour, painted in what is now western Wyoming, was the result of a commission from William T. Walters to document the overland trail. Miller (1810–1874) knew that distant water was a mirage when the horses did not 'quicken their motion, or snort'.

('The-first-thunder-that-falls'), a hunter of Omaw-whaw origin, finds love with the beautiful Zecana ('The Bird'). They live happily, subsisting on the wild goats of the Oregon Highlands and beaver from nearby glens, until a Spanish trader on the Mexican border persuades Ta-in-ga-ro to go pelt hunting at great distance and to leave Zecana behind. In the 'inmost recesses'[12] of the Rockies Ta-in-ga-ro finds the 'Boiling Spring', a magnificent fountain, 'the first you meet with after crossing the Great Desert', which is treated with reverence by Native Americans. Ta-in-ga-ro drinks fully and then offers beads 'to the divinity of the place'. To his dismay, these float back up to the surface, and are swept away from the gurgling spring: 'the soul of the intrepid savage sank within him as he beheld the strange reception of his reverential rite.'

Ta-in-ga-ro is gripped with a terrible sense of foreboding: 'A strange mist swam before his dizzied sight, and he saw . . . the reproachful countenance of Zecana reflected in the mysterious pool.'[13] Ta-in-ga-ro immediately returns in search of Zecana, who is traumatized and incoherent, 'hover[ing] near some revelation too horrible even to pass the lips of insanity'.[14] Zecana stabs herself shouting the name of the Spaniard, who Ta-in-ga-ro then kidnaps and, having stripped him, in a surprising – and disturbing –turn straps him to the corpse of his beloved Zecana. The anonymous narrator relishes the gruesome 'hideous compact'. 'Trunk for trunk, and limb for limb, he was lashed to his horrible companion.' Bound together with his victim, the horse rides into the Great Desert, where 'the fierce sunbeams, unmitigated by shade or vapour, fell with scorching heat on the disrobed person of the Spaniard, while the moisture that rolled from his naked body seemed to mould him more intimately into the embraces of the corpse to which he was bound'. Night-time dew 'seemed only to hasten the corruption to which he was linked in such frightful compact'. Ta-in-ga-ro drives him ever forward, allowing him small amounts of water to ensure the prolongation of the agony. Some claim, our narrator concludes, that Ta-in-ga-ro has retired to a quite life with the Cheyennes. Others believe that he leads a 'predatory band' of the Blackfeet. Still others 'aver that when the GHOST-RIDERS are abroad, the grim phantom of the savage warrior may be seen chasing them over the interminable wastes of the GREAT AMERICAN DESERT'.[15]

Ambrose Bierce's short account of mirage sightings in North Platte County, Missouri, in 1887 conjures a space very similar to that in Miller's atmospheric painting. One of the 'best known tricks of the mirage' involves 'overlaying a dry landscape with ponds and lakes'. But the 'magicians of the air' also produced memorably sinister superior mirages. One was composed of 'the most formidable looking monsters that the imagination ever conceived'. Taller than trees, 'the elements of nature seemed so fantastically and discordantly confused and blended, compounded, too, with architectural and mechanical details, that they partook of the triple character of animals, houses and machines'. Bierce then describes a hideous hybrid monstrosity,

> inextricably interblended and superposed – a man's head and shoulders blazoned on the side of an animal; a wheel with legs for spokes rolling along the creature's back . . . obscuring here an anatomical horror and disclosing there a mechanical nightmare . . . the whole apparition had so shadowy and spectral a look that the terror it inspired was itself vague and indefinite, like the terror of a dream.[16]

North Dakota was the scene of a later legendary Fata Morgana on the occasion of George Custer's departure from Fort Abraham Lincoln for Little Big Horn in 1876. The date was 17 May, and the Seventh Cavalry, with 150 heavy wagons, headed west in humid conditions. At this point, the cavalry could be seen marching in the sky upside down, ghost riders in the sky, prompting the wives of the Arikara scouts who accompanied Custer to wail lamentations.[17] Rumours of this event were recently questioned on a Native American web forum, 'Little Big Horn Associates Message Board', by one 'Sergeant' self-identifying as 'Elk Warrior', who asked:

> I have often heard that when the men were leaving Fort Abraham Lincoln in May of '76, Libby Custer observed a mirage – a reflection of the men in the sky above them. I have never heard of anyone else seeing that one, nor any kind of mirage in the sky besides that. Does such a thing

> as this exist? Do the skies in our Western U.S. sometimes, perhaps under specific conditions, show a reflection like that? I am tempted to ignore this 'vision' Libby had as a way to add to her husband's legend. But I've never seen the Montana sky. Is such a thing possible?[18]

The arid, more southerly parts of North America also host a significant number of Fata Morganas. One from Arizona was the subject of a memorable painting by the early twentieth-century landscape artist Harold Betts (illus. 16). About fifty years later, James Gordon, a septuagenarian Arizona weatherman, having decided that mirages – often of towns looming across the Mexico border – demanded empirical study, would sit atop Black Hill near his home town of Yuma with a camp chair, drawing board and binoculars.[19]

'Hot' mirages generally invoke spaces marked by extremity or threat. Sometimes rising temperatures encourage a memorable poetry of deliquescence. An anonymous correspondent in *Chambers' Edinburgh Journal* in 1836 described a mirage in Madras in which

> the whole landscape appeared to have given way, like molten silver . . . The buildings in the distance looked as if their foundations had been removed, while the shattered and broken walls danced to and fro, as if under the influence of some magical principles of attraction and repulsion; whilst many patches of imaginary water – the celebrated 'mirage' of the desert – floating where no water could have existed, mocked our sight in this fantastic landscape.[20]

Mirages in hot environments were ultimately cruelly dishonest, but while the delusion lasted, they were able to deliver cooling relief.

16 *Mirage on the Arizona Desert*, postcard after a painting by Harold Betts, *c.* 1915. The verso declares: 'In the deserts of Arizona and California the mirage is often seen, and in some locations it is visible nearly every day of the year. Broad lakes, surrounded by fields of green, herds of cattle, houses, and tree-bordered streams frequently appear and many a prospector or adventurer has been misled by the apparent oases, only at last to find that the spreading trees and the sparkling water evade him, the valley that looked so green is as dry and barren as the trail he left, and that he has been in pursuit of a mirage.'

This was beautifully expressed by the American poet George William Curtis in his *The Wanderer in Syria* of 1852. He conjured the 'cool lakes and green valleys' of the mirage that appears to the 'dying Bedoueen' where a

> voice of running water sings through your memory . . . the laughter of boys bathing there, – yourself a boy, yourself plunging in the deep, dark coolness – and so, wearied and fevered in the desert of Arabia, you are overflowed by the memory of your youth . . .[21]

Even more lyrical was W. H. Bartlett's evocation in 1848 of the landscape between Cairo and Petra. 'Half-dozing, half-dreaming,' he is lulled into a reverie:

> the startling mirage, shifting with magic play, expands in gleaming blue lakes, whose cool borders are adorned with waving groves, and on whose shining banks the mimic waves, with wonderful illusion, break in long glittering lines of transparent water – bright, fresh water, so different from the leathery decoction of the zemzemia.[22]

This last word, 'zemzemia,' refers to a leather water bottle which takes its name from the Zem Zem, the sacred well at Mecca. In the contrast between its 'leathery decoction' (a concentrated liquor) with fresh, transparent water, mirage appears as primal, natural and pure, something (although imaginary) that is positively contrasted with Islam.

Hot mirages can also be admired because of their superior aesthetics. An account of marine mirages praised Scoresby's Greenland sightings, and the inverted appearance of the entire English fleet in the Baltic in 1854, but then noted that 'it is in tropical seas that the most remarkable instances have occurred,' going on to record appearances in Mauritius and Bombay in which looming ships were seen from a distance of two hundred miles.[23]

More commonly, however, accounts of hot mirages often mixed anxiety about their compelling deceptiveness with a sense of danger and cultural disturbance. The greatest account of 'hot' mirages is provided by James Richardson's often harrowing account of journeys through the Sahara in the 1840s. Not without reason did a correspondent in the *Edinburgh Review* call him 'poor Richardson.'[24] In *Travels in the Great Desert of Sahara*, Richardson gives vivid testimony of the effects of heat and thirst on travellers close to death. Mirages appear as pathological interventions, distorting a visual field whose navigation is essential to survival:

> Every tuft of grass, every bush, every little mound of earth, shaped itself into a camel, a man, a sheep, a something living and moving... I was staggered at the deceptions and phantasms of The Desert. Every moment a camel loomed in sight, which was no camel. There was also a hideous sameness! The reason, indeed, I was lost.

On his way back to Misrata, in extremis, Richardson makes a fascinating connection between mirage and jinns:

> our brains reeled, and we all suffered from thirst. People seemed all mad to-day [a] camel-driver pretended he heard sweet melodious sounds. On inquiring what music it was, he replied, 'Like the Turkish band.' Then another came running to me, 'Yâkob, see what a beautiful sight.' I turned to look, but my eyes were so weak and strained, that I could see nothing upon the dreary face of the limitless plain. Essnousee swore to seeing a bright city of the Genii, and actually counted the number of the palaces and the palms.[25]

Richardson himself was too ill and weak to make out the mirage. However, the extreme shape-shifting, marking a breakdown of ordinary vision, was, as Richardson's account itself made clear, potentially the source of pleasure. For his camel driver, it was a beautiful sight. One year later *Chambers' Edinburgh Journal* reproduced an anonymous account of an 'enchanted' mirage from the *Cape Town Mirror*. Ostensibly a warning to other sailors approaching Walwich Bay (now Walvis Bay in Namibia), it narrates the crew's shock on seeing features of the coastal landscape that no one could recollect having seen before. Imagining that they had drifted off course, they then see a woman in a white shawl walking on one of the newly appeared islands. Then, half a mile distant, they spot a village inhabited by people in clothing of different colours: 'little naked brown children could also be distinguished running about at the edge of the water.' The narrator then leaps into the water, his splash producing a miraculous effect: 'the whole crowd of people on the shore great and small, gray and red, and brown, instantly soared up into the air and flew away in a cloud of pelicans, flamingoes, sand-pipers, and other birds.'

The disenchantment continues as the narrator advances up the shore, the 'village' revealing itself to be 'the skeleton of an enormous whale whose arching ribs had taken the appearance of a row of native huts'. It was very 'singular', the account concludes, and they were all taken in thoroughly by 'variations in the density and refractive power of the atmosphere'.[26] The image this conjures is suggestive of Salvador Dalí's *Paranoiac Face* of *c.* 1935 which turned a photograph of small (African?) figures seated in front of a domed dwelling, when aligned vertically, into a face (of the Marquis de Sade, according to Breton).[27]

## SIX

# Mirage and Crisis

Mirage often has a politics: it is rarely if ever only the product of atmospheric optics. The narratives that attach to mirages and the exact nature of what beholders believe they can see reflect, in part, the concerns and anxieties of their times. Times of crisis are prone to producing omens visible in the sky that have their origins as Fata Morganas. There are many ancient precedents. It is widely assumed that the Romano-Jewish scholar Flavius Josephus (37–100 CE) may have been describing mirages when he wrote of apparitions in the sky. Omens appear in the lead-up to Passover. In the most elaborate, 'chariots and troops of soldiers in their armour were seen running among the clouds, and surrounding . . . cities.'[1]

The English interregnum, the period immediately following the execution of Charles I in 1649 and the Restoration of rule by Charles II in 1661, was likewise a time of exceptional anomalous phenomena, some of which may well have been (and conversely may not have been) Fata Morganas. Ignorance of atmospheric refraction doubtless transformed many mirages into other phenomena, but we should be cautious in our diagnosis of what may have been visions driven by psychic fervour as misdiagnosed mirages.

The sense of guilt induced by the execution of Charles I would snowball into calls from Restoration clerics for repentance for the sinful regicide through prayer. Well into the eighteenth century the *Book of Common Prayer* detailed a prayer and fast to be observed every 13 January to

> implore the Mercy of God, that neither the guilt of that sacred and innocent Blood, nor those of other Sins, by which God was provoked to deliver both us and our King into the hands of cruel and unreasonable Men, may at any time hereafter be visited upon us, or our Posterity.[2]

The many monstrous births proclaimed by excited pamphlets, many of which were collected by George Thomason, a friend of John Milton, were anomalous aerial and optical phenomena. One pamphlet, *An Exact Relation of Severall Strange and Miraculous Sights Seen in the Air*, reported news from Hamburg and Dresden in early 1661. Without parallel in Europe, the text claimed, were the three ships which appeared upon the River Elbe,

> sayling with a full wind through [a] Bridge, and thence upon a suddain they appeared on dry Land; After which, immediately there came out of the first ship little Children, out of the second came forth men with naked Swords and Spades, hacking and hewing as they went, and out of the third came all Souldiers, and after they had been upon the Land, they with the three Ships suddainly vanished away.[3]

In the same year there were reports from Danzig that 'for several daies together there [were] seen in the Heavens many wonderful strange and miraculous Sights, as of Pikes, *Greek* and *Hebrew* Letters, and also in the night season great Armies of men in the habit of *Turkes*'.[4]

While it would be simplistic to see these merely as misunderstood mirages, there are nevertheless clear indications that atmospheric refraction may explain specific features of such powerful apparitions.

## SEVEN

# Oriental Mirages and 'Spectatorial Democracy'

The history of Oriental mirages, it is probably true to say, was inaugurated during Napoleon's Egyptian invasion of 1798–1801. It is widely believed that we owe the word 'mirage' (via the French, from the Latin *marari*, 'to wonder') itself to this 'expedition'. One of the many experts involved in this grand scheme, the mathematician Gaspard Monge, saw a standard inferior mirage: a body of alluring water that disappeared and retreated before reappearing. He shared in the disappointment of his thirsty colleagues but went on to publish a paper in the *Memoires sur l'Égypte* in 1800, this widely being considered the first optical theory of mirages. Monge describes marching through the desert between Alexandria and Cairo and experiencing mirage daily: the land was 'terminated, to within the distance of about a league, by a general inundation', turning villages into islands.

Monge himself actually suggests that the term mirage was already in use, for he notes that the phenomenon 'is familiar to mariners, who observe it frequently at sea, and have given it the name of *mirage*'.[1] It is this name that he chooses to use for marine and land-based appearances: although they may be quite different, 'the effect being the same in both cases, I have not deemed it proper to introduce a new word'.[2]

Hartwig's 1875 compendium of magical and sublime atmospheric phenomena positioned Monge not only at the beginning of the linguistic

17 John Tenniel, 'Napoleon's Army in Egypt Deceived by the Mirage of the Desert', illustration from W. Haig Miller, *Mirage of Life* (*c.* 1890).

recognition of what had been previously difficult to name,[3] but at the heart of the experiences, seductions and frustrations that travel writing and film have inserted deeply into our ongoing visual imagination.

> The illusion of the mirage was often a source of grievous disappointment to the army. When after a long march through the desert the deceptive horizon showed them the blue mirror of the lake in the midst of the arid sands, they hailed it with exclamations of delight, and hastened towards the imaginary shore; but, as they approached, it still receded before them, as if in mockery of their vain efforts to attain it.[4]

Monge was chiefly concerned to explain the optics of mirages, but he also comments on its military implications, noting that mirage produces illusions 'against which it is proper to be on our guard in a desert that may be occupied by an enemy'.[5] The appearance of a mirage in front of Napoleon's army was the subject of a perhaps fanciful John Tenniel drawing showing the parched French army struck by a 'gleam of hope' provided by an 'enchanted spot', a lake that 'appeared in the wilderness, with villages and palm trees clearly reflected on its glassy surface' (illus. 17).[6] Mirages appear as a military threat in the lead-up to the Battle of Alexandria in 1801. One British account notes how troops 'were prevented from advancing, and taking up an advantageous position, by imagining they were on the confines of a lake, or sheet of water'. This delusion persisted until 'the French were observed descending and marching across the imaginary lake to attack our regiments in the front. This ignorance occasioned a severe and needless loss of lives; the men being obliged to fight, under every possible disadvantage.'[7]

18 'Mirage on the Steppes of Central Asia', *Illustrated London News* (8 December 1888). Published purely for its visual impact, this image powerfully conflates Islam with mirage. The haggard rider in the foreground, together with animal bones on the right, and the scavenging vulture, add to the sense of impending doom.

Monge had experienced a typical inferior mirage (the phantom oasis) of the kind that would litter subsequent travel narratives. Rather than the ornate, fanciful and deeply seductive mystery of a Fata Morgana, inferior mirages were an almost routine test of any expedition, and took on some of the moral connotations that we have seen in the context of the Indian *mrigtrishna*. Frequently these inferior mirages provoke not the astonishment precipitated by elaborate Fata Morganas, but, as we have seen, a kind of internalized moral reproach, because the beholder permitted themself to fall victim to the hoax: they reveal themself to be susceptible to the mirage's 'mockery'. There are some exceptions: Charles Doughty in his celebrated *Travels in Arabia Deserta* (1888) is, well, doughty in his scepticism: 'Oftentimes in the forenoons, I saw a mirage over the flint plains; within my experience, none could mistake the Arabian desert mirage for water.'[8] (illus. 18)

T. E. Lawrence, who would write the introduction to a later edition of Doughty's *Arabia Deserta*, produced one of the greatest monuments to mirage as occlusion, an opacity that is also allowed a certain productivity. Lawrence's *Seven Pillars of Wisdom* presents mirage as an ever-present feature of the desert. Mirage first appears in his narrative mixed with desert heat as a silencing sword: arriving at the Red Sea port of Jidda (Jeddah), he describes 'the white town hung between the blazing sky and its reflection in the mirage which swept and rolled over the wide lagoon, then the heat of Arabia came out like a drawn sword and struck us speechless'.[9] More predictably, mirage blinds: he writes of those remaining 'minutes of the day in which the mirage did not make eyes and glasses useless'.[10] This occlusion acquires a positive quality in the vicinity of the old castle at Azrak where 'the mirage blotted its limits for us with blurs of steely blue which were the tamarisk bounds raised high in the air and smoothed by heat-vapour'.[11] When Lawrence arrives in Maan, on the Hejaz railway, occlusion is sutured with Ottoman sleepiness: 'for concealment we trusted to the mirage and midday drowsiness of the Turks.'[12]

This connection between mirage and Turkish dreaminess is something we will encounter again shortly. It alerts us to the manner in which these travel accounts, with their fixation with mirage, contributed to the enduring imagination of the 'Orient' as 'a phantasm', a space of occlusion and illusion linked to bad politics, as Edward Said would later argue.[13] Mirages as anomalous optical phenomena demand contextualization within a political history of light.[14] Within this broad field, the deception of mirage raises the issue of transparency. Transparency and visibility remain fundamental to Euro-American political discourse. World governments are evaluated and ranked by the anti-corruption NGO Transparency International (their website map of perceptions of corruption could almost serve as an illustration

19 Trade card for the Amsterdam cocoa and chocolate firm Van Houten, late nineteenth century. In this remarkably 'Orientalizing' image, a city composed of mosques and minarets rises from the desert, to the astonishment of the beholders in the foreground.

of Said's thesis) and Norman Foster's design for the Bundestag dome on top of Berlin's Reichstag created a canopy of glass and steel in order to monumentalize the necessity of inspection and clarity.

These ideas feel very much part of our current epoch and yet they have a long and intriguing history. Perhaps the clearest statement of them was delivered between 1835 and 1840 by Alexis de Tocqueville in his work *Democracy in America* and has since become known as the 'spectatorial mode of politics'.[15] Democratic citizens, Tocqueville wrote,

> like to observe very clearly the object they are dealing with; they remove it, therefore, as far as possible from its envelope, brush aside everything which keeps them apart from it, and remove everything which hides it from view, so as to see it close to and in full light. This proclivity of mind soon leads them to despise outer forms which they consider as

> a useless and inconvenient veil positioned between them and the truth.[16]

The hypothesis I propose here is that mirages assume particular prominence in Orientalist narratives when they are linked to 'bad' politics, and especially to what became known as Oriental or Asiatic despotism. The experience of mirages was not, of course, uniquely an artefact of Orientalist description. Mirages are actual physical phenomena resulting from heat gradients, they are universally (optically) perceived in ways that are not subject in any marked way to cultural conditioning, and they can be photographed. Although they may well not be 'true', they are undoubtedly 'real'. Consequently humans, in different parts of the world, have marvelled over the millennia at these remarkable visions. It seems that all humans 'see' mirages in similar ways (while inflected by interpretive schemata), but of course they have produced different theories to explain them. And, more importantly for the specific hypothesis proposed here, there are parts of Asia where mirages have been commonly experienced but where, because they did not coincide with an anxiety about Oriental despotism (and especially the role of Islam within this), they have never been systematically 'Orientalized' in the ways that they have been elsewhere.

EIGHT

# From Clam-monsters to Representative Democracy

Mirages have been and are part of Japanese life. One of the most extensive Internet resources of the optics of Fata Morganas documents optical phenomena on and around Lake Biwa, just north of Kyoto.[1] Furthermore, the mirage is woven into Japanese mythology through the image of the island of Horai, known to the Chinese as Penglai. The zone of transculturation between China and Japan of which Penglai/Horai was one product and Fusang (the mysterious land lying to the east of China) was another has been the subject of a wonderfully inspired discussion by Edward H. Schafer. Fascinated by the 'pavilions, eyries, and follies that occasionally drift by, or materialize before the astonished eyes of early travellers in the open sea', he finds himself enquiring into seventh- to tenth-century poetry, especially that addressed to Chinese dignitaries heading east, for clues. The poet Ch'ien Ch'i (722–780 CE), for instance, predicts that an ambassador to Japan will encounter 'two phantom paradises hovering over the waves' in which will be seen 'the high houses of the clam-monsters bannered with rainbows'.[2]

The clam-monsters were prodigious molluscs that Chinese and Japanese belief saw as the origin of Fata Morganas, these 'phantom paradises'. The exhalations from these clams 'sometimes burst the film of surface tension and appeared to astonished mariners as stunning mansions adrift on the surface of the deep'.[3] In 1907 Lafcadio Hearn, the Greek-Irish Buddhist convert, lecturer in English Literature at the Imperial University of Tokyo from 1896 to 1903 and incomparable fabulist and romancer of all things Japanese, would publish *The Romance of the Milky Way*. This included a short section concerning the mythology relating to *shinkiro* (mirage), noting that images sometimes showed a toad exhaling a vapour in which was figured the island of Horai, but that even more common were images which (echoing the poetry that Schafer discusses) depicted a clam or mollusc (*hamaguri*) which 'sends into the air a purplish misty breath, and that mist takes form and defines, in tints of mother-of-pearl, the luminous vision of Horai'.[4] This legend is marvellously imaged in Kitagawa Utamaro's teacher Toriyama Sekien's supernatural bestiary of 1781, *Gazu Hyakki Yagyo* (The Illustrated Night Parade of a Hundred Demons). A double-page woodblock print in the first volume of the third book depicts a huge clam exhaling a wonderfully ornate Fata Morgana in which the Dragon King's Palace nestles among the mountains of Horai. The same molluscular birth of a mirage is

20 Utagawa Kunisada II, *Mirage in Spring Mist* (*Harugasumi shinkiro*), 1863, colour woodblock triptych.

21 Utagawa Kunisada II, *A Courtesan by the Shore Surprised by the Appearance of a Mirage*, c. 1854–60, colour woodblock print.

presented in Shunsen's *The Clam's Mirage of the Dragon Palace* of 1830, in which a delighted group on the shoreline point at the miraculous vision.[5] Beachside spectators are also deployed in Utagawa Kunisada's beautiful colour woodblock triptych *Mirage in the Spring Mist* (1863), in which a female clam seller points to a city of temples visible in a mirage exhaled by a giant clam (illus. 20). In the same artist's earlier fan print, *A Courtesan by the Shore Surprised by the Appearance of a Mirage* (*c.* 1854–60), we see a 'modern beauty' looking to her right at a clam from which a mysterious small crowd is emerging (illus. 21).

Both Sekien's and Kunisada's images bring with them the odour of *sfumato*, the wispy vapour of the clams turned into the smoky by-product of some magical transformation. Although the biology of mirage production here appears powerfully local and Japanese, the linkage between mirages and exhalations was common in early

European travel narratives. The cleric and traveller Thomas Shaw, writing about Barbary and the Levant in 1738, for instance, noted the miragistic quivering and undulating motion produced by a 'quick Succession of Vapours and Exhalations'.[6] Schafer, significantly, suggests that the derelict word 'fnast' (which the OED defines as 'breath', especially an exhortation or snort), has two advantages as a descriptor for this *sfumato*, one transcultural and one empirically descriptive: it is not 'encumbered' by assumptions about concepts such as 'breath' and 'pneuma' and is furthermore 'consonant with the bubbly wheezing of giant clams'.[7]

The connection between clams and mirages can be found in unexpected places. A postcard of 1913 shows a 'superb shinkiro (mirage)' and is marked on the verso with a souvenir stamp from an aquarium which encloses the outline of the entrance to the aquarium in the shape of a clam (illus. 22 and 23).[8] Actual clamshells were commonly decorated with elaborate painted scenes and used in games of shell-matching (*kai-awase*) and housed in large collections (up to 360 paired shells) kept in lacquered containers (*makie kaioke*). Clams also feature prominently in the magnificent carved decorations in the seventeenth-century Toshogu Shrine in Tokyo's Ueno Park, where they appear on the cusps of ocean fringes, thrown to the top of breaking waves, where they occupy the liminal space between water and vapour.

However, European Orientalist enthusiasm for Japan has rarely generated mirage literature, and mirages do not seem to be the source of cultural or political anxiety. One of the few further European commentaries on Japanese mirages also comes from Lafcadio Hearn. A short story, called 'Horai', opens with a memorable and quintessentially Hearnian pellucid eulogy to a picture of something vanishing. Hearn is gazing at a kakemono, a painting on silk, of a *shinkiro* (mirage), which is also a depiction of the island of Horai:

22 and 23 Japanese postcard depicting a *shinkiro* or mirage, Taisho 2 (1913). The verso is marked with a souvenir stamp from an aquarium, in the shape of a clam from which mirages were often shown emerging.

Blue vision of depth lost in height – sea and sky interblending through luminous haze. The day is of spring, and the hour morning . . . Only sky and sea, – one azure enormity . . . In the fore, ripples are catching a silvery light, and

24 Kojima Shogetsu, *The Imperial Diet*, Tokyo, 1890, colour woodblock triptych. This wonderful depiction of spectatorial democracy, and of the emergence of a new political transparency, helps explain why Japanese mirages were never 'Orientalized'.

> threads of foam are swirling. But a little further off no motion is visible, nor anything save color: dim warm blue of water widening away to melt into blue of air. Horizon there is none: only distance soaring into space, – infinite concavity hollowing before you, and hugely arching above you, – the color deepening with the height.[9]

Hearn then describes the shadowy outline of the Palace of the Dragon King 'far in the midway-blue' before lamenting the transience of this vision:

> Evil winds from the West are blowing over Horai; and the magical atmosphere, alas! is shrinking away before them. It lingers now in patches only, and bands – like those long bright bands of cloud that train across the landscapes of Japanese painters. Under these shreds of the elfish vapor you still can find Horai – but not everywhere ... Remember that Horai is also called Shinkiro, which signifies Mirage – the

> Vision of the Intangible. And the Vision is fading – never again to appear save in pictures and poems and dreams . . .[10]

Hearn fuses the East with colour, illusion and transience.[11] The 'evil winds from the West' are incompatible with Horai. By the time Hearn was writing (in 1903–4), much of what Hearn loved about Japan was fast receding into the past, recoverable only, as he says, in pictures, poems and dreams. The Imperial Diet had opened in 1890, and Hearn's story appeared in his book *Kwaidan: Stories and Studies of Strange Things* at the very moment that a Westernized Japan ('girding itself with [a] Western energy of will', as the anonymous introduction to *Kwaidan* put it) was preparing to fight Russia.

Japan had inaugurated its own version of spectatorial democracy in 1890 with the establishment (under the Meiji Constitution) of the Imperial Diet, and it was this perhaps that consigned the *shinkiro* to the past. 'Meiji', was written with two ideographs denoting 'bright government', not the opacity of archaic *sfumato. Bunmei kaika* ('civilization and enlightenment') was to be found not only through technological modernity but, so the champion of Western values Fukuzawa Yukichi argued, through a liberal spirit of free enquiry and dispute.[12] The Imperial Diet, despite its limited powers, embodied the importance of representative democracy alongside a constitutional monarchy, a duality evidenced in many woodblock triptychs from the period. In most of these the Emperor can be seen seated high up, looking down on the proceedings,[13] and sometimes (as in the case of illus. 24) he is magnified in stature, giving corporeal substance to his constitutional superiority. But all these images share a striking architecture of spectatorship in which a central figure addresses members of the Diet as if they were the audience in a theatre, speaking through the fourth wall in an unveiled panopticon: everyone sees everyone else.

## NINE

# The Halted Viewer and *Sfumato*

Kunisada's image (illus. 21) dramatically foregrounds a viewer of the mirage: in his case a 'modern beauty'. Mirages are frequently presented as phenomena for beholding, but as part of an image in which someone else is already doing a significant part of the beholding. One of the reasons for this may well have been the difficulty, before lithography and colour photography, of actually seeing the mirage in mass-reproduced images. The internalized beholder may have been necessary in order to direct the viewer's attention to an inadequately presenced mystery. Thomas William Atkinson, quoted in 'The Amoor and the Steppes' in *Harper's Monthly Magazine* in 1860, gets to the heart of mirage's quixotic representability: 'I fear my pencil fails in rendering [mirage's] magical effect, and my pen cannot give an adequate idea of its tantalizing power on the thirsty traveler.'[1]

Joseph Leo Koerner, in his virtuoso examination of Caspar David Friedrich, argues that the *Rückenfigur*, the halted traveller, serves for Friedrich as a figure of belatedness who obstructs our possession of the picture space.[2] In the absence of the halted traveller, our eye moves freely through the landscape represented within the image. The presence of that figure 'radically alters the way we see the painted world beyond us . . . I do not stand at the threshold where the scene opens up, but at the point of exclusion, where the world stands complete without me.'[3]

Koerner's argument is powerful and compelling but in the case of figures within mirage images, the position that Koerner argues against (in respect of Friedrich) more usefully illuminates their function. Many of Friedrich's images bear a resemblance to Jan Luiken's illustrations for Christophoro Weigelio's emblem book *Ethica naturalis* of 1700, in which the figure 'will gesture towards the scene, as if to say, "Behold!"'[4] This is the kind of conventional framing that Koerner rejects but which seems to explain the function of *staffage* (subsidiary human figures) in mirage images.

The halted viewer, in mirage images, fulfils a function that was often dictated by the difficulty that particular media had in representing mirages. The lone camel rider in the Libyan desert serves as an aid to our perception of what can hardly be seen (illus. 25). We are asked to follow the vision of these internalized spectators halted in astonishment at what we as secondary viewers can hardly imagine. Often this internalized *staffage* points to what would otherwise be overlooked. In Japan the mirage was explicitly seen as a smoky *sfumato* whose identity, allied as it was to the clam from which it appeared, was rarely in doubt. Within European representation, and especially within the limitations of pre-photographic mechanical reproduction, the mirage had a difficult and uncertain valence.

Michael Gaudio has drawn attention to the connections between the depictions of smoke, witchcraft and magic in Theodor de Bry's late sixteenth-century engravings of the Americas. This was a connection that can be sustained across a range of images such as Hans Baldung's 1510 woodcut of a *Witches' Sabbath* and Martin de Vos's later *Saturn and his Children*. However, Gaudio also focuses on the actual traces in de Bry's engravings through which smoke is made visible ('visible traces of the engraver's hand and of the difficult process of pushing the burin into a resistant metal plate'[5]). The 'roiling, spiraling smoke' that features in many of de Bry's images of Native Americans rises not only 'toward the sky but also toward the surface of the page',[6] suggesting the dangerous nature of mimesis itself. Representation becomes a kind of smoke, prompting Gaudio's use of the term *sfumato*. Used to describe Leonardo's legendary 'smokiness', his ability to magically fuse matter with air 'by causing forms literally to merge with their surrounding atmosphere',[7] it captures also the illusionistic seduction and danger of mirage. Into this can be folded the interpretive dimension of mirage. Though optically real, mirages operate through a form of gestalt. The founder of American cultural anthropology, Franz Boas, in the process of switching his focus from physics to ethnology, had travelled to Baffin Island in the 1880s to study local perceptions of the colour of seawater. Although he was very interested in refraction, and documented mirages, he was unable to 'localize' mirages or to show they were subject to 'apperception', that is, subject to a 'cultural

25 *Libyan Desert, Mirage on the Horizon*, steel engraving after a photograph by M. D. Heron, late 19th century.

eye' that somehow saw them differently in any radical way. And yet if we look at lithographs of Fata Morganas (such as Abbott's – see illus. 1 and 2) and contrast them with photographic documentation, it is clear that the basic optical building block of mirage is enormously elaborated by specific cultural imaginations. In the case of Abbott, we can see that what Scoresby termed the 'columnar' threads of the Fata Morgana become the starting point for the transformation of inverted elements into ships in full sail and mosques and minarets (see the top left of illus. 1). Mirages offered partially filled tabulae rasae on which 'Ulysses' sail' could be unfurled.[8] Their otherness and obscurity invite fantasies of the exotic – of ocean-going vessels and Eastern devotion. A hundred years later, the retired Arizona meteorologist James Gordon would make a similar observation about a mirage seen, a century before but still remembered, in a small unnamed Maryland town. 'It showed a city in the sky, a city of domed roofs – foreign looking, not at all American. Judging from appearances, it would seem to have started on its long journey from North Africa.'[9] Japanese mirages, by contrast, always feature Japanese architectural styles.

The visibility of the mirage becomes a question not simply for human perception but for the mirage in the age of mechanical reproduction.[10] Woodblocks seem to conjure mirages more easily than steel engravings, which had particular problems with inferior mirages, but it is lithography (as we have already seen with Abbott in central India) and chromolithography that really excel in projecting the mysterious qualities of both inferior and superior mirages. Photography offered a technological autopticism or eyewitnessing that fixed and affirmed these ultimately slippery phenomena, but the details could be more easily brought out through engravings than early half-tone photographic reproductions.[11] Thus *The Graphic* presented what it suggested was 'the first instance of a mirage having ever been photographed'. The image,

26 Orr & Barton, 'A Curious Mirage Recently Photographed at Madras'. Captured with a camera in June 1885 by the well-known Bangalore photographers, this image of a Massula boat was presented by the British weekly illustrated newspaper *The Graphic* as 'the first instance of a mirage having ever been photographed'.

'taken quite by chance' by the celebrated Bangalore photographers Orr & Barton, shows a Massula boat photographed near the pier at Madras on 17 June 1885. The 'engraving' (*The Graphic* does not specify the exact technique) emphasized the Fata Morgana, which projected the boat and larger distant steamer into the sky. The image took the authority of the photograph's 'index' (its relationship of casual continuity to what it depicted) and blended it with the engraver's ability to emphasize and accentuate the most fascinating elements of the image (illus. 26). The Japanese postcard from 1917 mentioned above, by contrast, presents the distant blurred presence of a massive Fata Morgana to much less impressive effect.

27 R. G. Willoughby, *R. G. Willoughby's Mirage (the Silent City) Alaska*, 1888, photograph.

Photography, it turns out, was imagined to be the redeemer of mirages almost from its very start. In 1846 *Scientific American* reported a Fata Morgana in Pomerania lasting twenty minutes. It was 'deep blue [in] color . . . on a brilliant opal-colored ground with extraordinary clearness and precision' and 'reflected with such exactness that it appeared to be a daguerreotype design'.[12] The following year, the same journal reported a 'Splendid Mirage in Paris' in which the spire of the Cathedral of Ulm appeared between seven and eight in the morning in an image 'so correct that it might be mistaken for a representation made by a daguerreotype'.[13] Inevitably there were competing claims for the earliest photograph of a mirage. One tantalizing account offered by a correspondent writing under the pen name 'Glatton' claimed to have captured a mirage with a camera in 1879. The intriguing image (which he says may still be available from the Tenby photographer Robert Symons) shows an inverted gunboat, subsequently identified as the *Gadfly* then being launched twelve miles away in Pembroke, halfway up the spire of a church in Tenby.[14] The anthropologist Franz Boas would photograph several mirages during his voyage to Baffin Bay in 1883–4.[15] A few years later, photographs purportedly of Alaskan mirages, made in 1888 and 1889, would feature in a fascinating episode.

One of the photographs, authored by Professor R. G. Willoughby, claimed to be a record of an astonishing mirage in the vicinity of the Muir Glacier in what is now Glacier Bay National Park (illus. 27). Retailing at 75 cents a copy, it showed a 'Silent City' emerging from the glacier. Alexander Badlam, the Mormon pioneer and author of *Wonders of Alaska*, was outraged when he encountered these 'glacial joke[s]' in his travels. He noted that no one doubted that mirages existed; indeed, he goes on to provide some remarkable descriptions of what must have been astonishing visions. However, Badlam could not help but see 'a reflection on the intelligence of the average mind when the public is requested to believe that the city of Bristol, England, has been photographed on top of the Muir Glacier'.[16] Willoughby is presented as a gullible backwoodsman (he had 'never seen a locomotive', for instance), and his superimposition of the city of Bristol in this unlikely Alaskan location is presented by Badlam as wholly risible.

But it got more bizarre. *The Daily Transcript*, a Nevada City newspaper, reported on the adventures of one James O'Dell, who had left California to work in a gold mine in Alaska. Being familiar with the 'Silent City', he set about trying to see it. In his earlier life prospecting in California, he perfected an almost magical device, worthy of the *Arabian Nights*, that gave forewarning of the approach of strangers. This involved placing a 'few pounds' of quicksilver (mercury) into a gold prospecting pan and then peering into it with a magnifying glass.

'In this way we could detect anything that moved on any road or in any place for miles around. The face of the country and all upon it was first reflected upon the heavens or upper strata of air, and thence upon the pan of quicksilver.'[17] O'Dell and a companion cruised around in front of the Muir Glacier for a day or two hoping to see the mirage, without success. They then decided to try divination by quicksilver and were immediately rewarded with an image of what appeared to be a large ruined city. They elaborate a fascinating American folk theory of mirage:

> We saw enough to convince us that the city was at the bottom of the bay, was thence imaged on the clouds and then reflected down upon the quicksilver. It may be that, in certain favorable stages of the weather, the image of the sunken city is thrown upon the glacier, where it resembles a mirage.

They then ascended the glacier (it took a whole day) and mounted a mirror on a tripod, at a height of 1.5 m (5 ft), in which they could also see the ruins of the city. 'We were not a scientific expedition,' they modestly concede, 'but in our own rough way we were able to satisfy ourselves that what is called the "Silent City" is in reality a sunken city resting at the bottom of Glacier Bay.'[18] Proof of this was then established by the photographer I. W. Taber (illus. 28).

It got more complicated. The San Francisco *Examiner* interviewed George Kershon, who in 1888 had travelled into the interior of Alaska. He had quite a tale. He had bought a sloop to get him up to the Yukon. Then he fell out with his partners and went by canoe with two Native Americans up an unknown fork of the river, where he had 'a terrible time'.[19] Two pages of icy mountainous tribulations later,

Kershon is greeted with an uncanny vision of the 'strangest thing', another phantom city. 'After several hours of hard work I reached the outskirts of this mysterious city, and found that the place was laid out in streets, with blocks of strange-looking buildings, what appeared to be mosques, towers, ports, etc. and every evidence of having been built by art.' Ghostly, and quieted by an 'awful stillness', the empty streets were often blocked by ice. Returning via Juneau, Kershon became aware of the Muir Glacier images and realized that he had encountered the same city: 'the mirage of Muir Glacier is the reflection of the frozen city found by me.'[20] How the ruins of a huge city came to be in Alaska he is unable to explain.

Photographic proofs of phantom cities were to cause problems for Badlam, who saw his own mirages and had his own camera. How was he to establish the authority of his own images? For a start, he stresses the community of beholders (the passengers of the steamer *Ancon*) who shared his vision of mirage. Eight to ten miles south of

28 I. W. Taber, 'Taber's "Silent City"', Glacier Bay, said to have been photographed in July 1889. Reproduced in Alexander Badlam's *The Wonders of Alaska* (1890).

Pacific Glacier he and his fellow passengers saw 'what seemed to be a block of large white buildings . . . Beautifully formed spires, apparently three or four hundred feet high reached above the buildings.' Badlam photographed this and reproduced this image in his book of 1890 in which it jostled against Professor Willoughby's and I. W. Taber's images. In this evidentiary competition, Badlam seeks the autoptic support of a deposition, a 'card' that 'proved the existence of a mirage'. Signed by two gentlemen, Robert Christie and Robert Patterson, it testified that 'we suddenly saw rising out against the side of the mountains what appeared to be houses, churches and other huge structures. It appeared to be a city of extensive proportion, perhaps 15,000 or 20,000 inhabitants.' They went on to stress that they had never seen Willoughby's photograph. In this manner, Badlam anxiously sought to protect plausible mirages from those which he considered only worthy of 'Baron Munchausen's fairy tales'.[21]

By 1898 a physicist from the University of Wisconsin was reporting in *Nature* on his experiments during summer visits to San Francisco photographing mirage effects on the city's flagstone sidewalks.[22] Success in capturing images of sidewalks 'flooded with a perfectly smooth sheet of water' was achieved through the use of a 'very fine tele-objective' lens. A few years later, *Scientific American* was reporting on the development of the 'Telephot', capable of capturing 'any phenomenon visible at the extreme horizon, such as mirages'.[23] Alfred Wegener's photographic records of Arctic mirages in the Danish Greenland expedition of 1906–8 were widely acclaimed as 'most remarkable'.[24] By 1913 stereoscopic photography, with its 'numerous advantages over ordinary photography', was thought ideally suited to depicting subjects, including mirages, which in ordinary photography 'appear insignificant'. Stereoscopy, it was believed, would give them 'a fresh and surprising charm when standing out in the relief of a stereoscopic photograph'.[25]

Mirages were delicate objects that were seen to require a photographic handmaiden to make them fully visible in durable form. But they also shared with photography a destabilizing magical quality. A fascinating discussion in *Scientific American* in 1892 (reporting an earlier piece in *La Nature*) opened with a commentary from one Paul Roy, professor at the Lyceum in Algiers, documenting the appearance of his own 'mirage' in a photograph he had taken of his son, and then went on to discuss an image described in the *Journal of the Photographic Society of India* in which a figure pictured in front of the Himalayas proved, on closer inspection, to be 'entirely transparent'. Stones and other objects on the snow could be seen through his arms and legs. This was a possible 'mirage' that was ultimately neutralized (like Professor Willoughby's work) as a 'double exposure', its effect being attributed to a small aperture in the camera caused by a missing screw.[26]

TEN

# Memory and Modernity

Mirages, both literal and metaphorical, frequently conveyed a negative image about what was obscure, illusory and deceptive.[1] But the motif also created a space of intense ambivalence where a magic that could never be destroyed was offered a disavowed reverence. Hence the mirage motif is rarely simply condemnatory. More commonly it conveys a sense of escape, of fascination, of a desire to be deceived and of sheer visual delight. This ambivalence perhaps reflects the ineradicable desire to be deceived, to become a victim (in part, willing) of the *sfumato*, or what M. R. James in his classic ghost story 'The Mezzotint' called the sinister and agentive 'sympathetic ink' of the image.[2] This desire to be deceived has very deep roots and can be traced back at least to Joseph Addison's praise in 1712 of the 'pleasures of the imagination' in *The Spectator* in which he declared that 'Things would make but a poor Appearance to the Eye, if we saw them only in their proper Figures and Motions,' noting that 'our Souls are at present delightfully lost and bewildered in a pleasing Delusion.'[3]

It was probably Monge who gave us the term 'mirage', but photography and film were later to prove responsible for the ubiquity of mirages. The year 1800 marked the moment that allowed mirages to be named, but it was late nineteenth-century developments in photography that allowed mirages actually to be seen, and twentieth-century film which helped them burrow inside our heads. In the early twenty-first century we have only to close our eyes to be able to conjure a mirage: heat haze shimmering on a hot tarmac road, or the seductive allure of a glistening desert oasis. These are fragments from a repertoire that we have internalized via technology's own answers to the mirage. Anyone who has seen David Lean's *Lawrence of Arabia* will carry somewhere in their subconscious a trace of Ali's appearance in the desert.[4] In a steadily held long shot, filmed by cinematographer Freddie Young with a special 482 mm Panavision lens, a magical optical space of anticipation and transformation conjures a distant shape-shifting horseman, Ali ibn el Kharish, who (in a witty inversion of the presence of water in a mirage) comes from afar to claim rights to a well from which Lawrence is drinking. The water is real; the beholder from afar is rendered under a possible erasure in a precise inversion of the conventional mirage. This memorable sequence transposes the usual relationship between observer and mirage as a thirst-quenching possibility. We understand at the beginning that the water is real, and the 'mirage' then also turns real when Ali shoots Lawrence's Bedouin guide.[5] Werner Herzog's *Fata Morgana* of 1971 is ostensibly a report on a dying planet by extraterrestrials. Shot in the Sahara and Sahel, it is in parts (as its title implies) preoccupied with the mirages that sparkle in the desert landscape. Inferior mirages occupy a zone on the horizon where mimesis pulsates between presence and absence. Oscillating vehicles limn the space between figuration and abstraction, performing a role similar to Herzog's camera.

This memory bank helps explain a puzzling outbreak, in the late nineteenth and early twentieth centuries, of mirage reports in scientific literature. In earlier times mirages were largely placed 'in nature' as exceptionally rare phenomena distant from humans and requiring specific techniques of memory to store them for future use. Julian R. Jackson, Secretary to the Royal Geographical Society, produced a 'remembrancer' as early as 1841, directing the traveller how and what to observe. His manual was designed to excite the traveller's curiosity but also to discipline it and to encourage the preservation of observations. Alongside hoar frost, falling stars and rainbows, Jackson also included a section on mirages that perhaps predictably drew on Monge, Humboldt and John Malcolm's Persian observations among others and mentions the Enchanted Island,[6] Cape Fly-away (the mirage-afflicted Amundsen Gulf)[7] and the *Flying Dutchman*. His short account is spiced with exotica such as Tod's account of a vertical mirage at Hissar, India (known locally as the City of Rajah Hirchend (Hari Chand), 'celebrated in the fabulous history of India'), but is also at pains to direct the remembering eye towards domestic appearances such as 'the village of Great Paxton with its dependencies', which was seen suspended in the sky on 16 July 1820.[8]

Later contributions, such as numerous short letters to the journal *Nature*, are characterized by a surprising sense of novelty, conveying a powerful feeling that modernity is only just starting to make mirages familiar. In 1870 Sydney Skertchly reported a 'vivid' experience on a hot, dry day in the Wash ('equal to any mirage I have witnessed on the African desert'). The River Nene appeared to be overflowing its banks, which he 'knew to be quite out of the question, but the semblance was so perfect that it required an effort to believe that it was but an illusion'.[9] In 1890, one Arthur E. Brown reports the experience two years earlier of seeing a 'small conical-shaped island . . . with its inverted image' in the Sea of Marmara off Constantinople before noting that 'I do not know whether mirages at sea are uncommon; but as the officers on board did not remember seeing one before, I thought these instances might be worth recording.'[10] In all these reports, the mirage is rare and exotic. In the twentieth century it would become domesticated and part of the everyday.

We have already encountered R. W. Wood's experiments with sidewalk mirages in San Francisco in the 1890s. By the 1920s, in the UK, the motorcar and metalled roads (tarmac was patented in 1901) had established new infrastructures for the production of mirages. A Midlands correspondent recorded the frequency of mirages on 'tarred macadam, or wood block', the optical effect being of 'a layer of water . . . immersed in which are the feet of pedestrians and the wheels of vehicles'. This was echoed by Harry Hilman, reporting in 1920, who saw his first mirage in Colombo but now witnessed them repeatedly in England, especially where there were 'tarred roads . . . bright sun; and . . . a slight gradient rising from the observer'.[11]

29 James Sayers, *Galante Show*, aquatint, published 6 May 1788. Edmund Burke is depicted as the manufacturer of mirages during the trial of Warren Hastings.

# ELEVEN

# Theatrical Mirages

As Yaron Ezrahi has noted, Tocqueville's association of democracy with transparency derived much of its power through the opposition of an 'honest' visible democratic politics with the 'dishonest theatrical politics of the monarchy and aristocracy'.[1] To this we might add the threat of what became known in Britain as 'Nabobism', the disruptive wealth that the East India trade in particular was bringing to Europe.

'Nabobism' was one of the accusations levelled at Warren Hastings, the first Governor General of India following his return to England in 1784. From 1788 to 1795, Hastings would find himself at the centre of impeachment proceedings for corruption, whose 'self-conscious theatricality', as Flood terms it in his wonderful account, perfectly suited the setting of Westminster Hall. Contemporary engravings depict 'Warren Hastings Esq. Prisoner' seated dead centre in front of the Commons Committee. On each side are the Counsel for the Prosecution and the Defence, and in tiered seating all around are hundreds of spectators. The whole arrangement, reminiscent of Centre Court at Wimbledon, was justice as theatre but also justice as visibility. At the heart of this, the prosecution, led by Edmund Burke, made a broader case about 'the intimate relationship between ways of seeing and governing' India.[2]

The issue of a 'true' as opposed to 'distorted' vision was central to much of the rhetorical battle within the trial, and the source of ingenious commentary in the popular visual culture that responded to it. Flood characterizes this as a 'battle of the lenses'. The playwright Richard Brinsley Sheridan, for instance, in a famous speech contrasted the 'deformed idol' advanced by Hastings ('this base caricature – this Indian pagod – this vile idol hewn from some rock – blasted in some unhallowed grove – formed by the hands of guilty and knavish tyranny to dupe the heart of ignorance') with the 'true image of justice'.[3]

The charge and counter-charges relating to the Begums of Oudh, in 1788, were the subject of vigorous and witty caricature in a series of aquatints and engravings. James Sayers published an aquatint on 6 May 1788 titled *Galante Show*, 'galante' describing an itinerant magic lanternist. In the image we see a bespectacled Edmund Burke manipulating a projector that throws up a magnified image of a Benares flea, a Begum's wart transformed into a series of mountains, and the Begum's giant eyeballs producing tears in which (taking its cue from Polonius) an 'Ouzle' becomes a whale (illus. 29). A few days later James Gillray responded with *Camera-obscura*, which inverts much of Sayers's design. In place of Burke we see Hastings, wearing the headdress of a 'Nabob', who uses his camera obscura to project a miniaturized view of events whose 'true' magnitude is shown at the top of the picture. An elephant becomes a flea, a mountain becomes a wart, atrocities perpetrated against Indians become 'Skind mice', and the whale is reduced to an Ouzle. Flood notes that 'Hastings's acculturated dress is more than a personal foible, hinting at moral corruption and accusations that he had bribed his accusers,' pointing the viewer towards a widespread anxiety about the transformation of class relations under the hammer blow of mercantile wealth and towards the idea of 'Asiatick despotism' (illus. 30).[4]

30 James Gillray, *Camera-obscura*, aquatint, published 9 May 1788, depicting Warren Hastings as a 'Nabob' minimizing the evidence against him. The events in India shown in their true proportion at the top of the image are greatly reduced in Hastings's projection at the bottom of the image.

Anxieties about Nabobism and the visual preoccupation with projection and distortion that features so explicitly in the Hastings images are also a recurrent theme in Thomas Rowlandson's illustrations in 1816 for William Combe's *The Grand Master, or Adventures of Qui Hi? In Hindostan*.[5] The plate titled 'Phantasmagoria: A View in Elephanta'[6] references Athanasius Kircher's famous image of a devil projected by a magic lantern and has two seated incarnations of 'avarice' and 'misery' viewing a magic lantern image in which a satanic bat is transformed by the projector into a gentleman's genteel wig. The familiar moral appears to be that visual trickery can persuade us that evil is capable of appearing 'civil'. Sayers's, Gillray's and Rowlandson's wonderful cartoons draw attention to the machinery of representation such as magic lanterns and the camera obscura, transforming the question of fidelity and trustworthiness into one of mediation. The matrix of nature, techne, and truth and illusion, had long been present through the influence of the Jesuit propagandist Athanasius Kircher.

For Sayers and Gillray, mediation is negative: mechanical artifice enables a wider dishonesty. By the 1920s, the German philosopher and writer Salomo Friedlaender was able to imagine cinema as a 'Fata Morgana machine'. In his short story of that name, he narrates the dreams of one Professor Abnossach Pschorr to 'achieve the optical reproduction of nature, art, and fantasy through a stereoscopic projection apparatus that would place its three-dimensional constructs into space without the aid of a projection screen'. One cannot help but wonder whether Friedlaender was familiar with the remarkable image of the Straits of Messina Fata Morgana that had appeared in *Die Wunder der Natur* encyclopaedia of 1913 (see illus. 5). The professor is successful and contrives a 'purely optical phantasmagoria' with the aid of 'stereoscopic double lenses' such that three-dimensional forms are 'detached' from the surface of the 'projection screen'. Friedlaender's story then relates the professor's attempt to sell his invention to the military (it is rejected on the grounds that it might bring an end to war) before providing a memorable description of a European mirage haunting the 'Orient':

> When a storm is brewing... it is unclear whether this storm is only optically real or a real one. Through and through ... even the Orient fell into confusion when a recent Fata Morgana produced by solely technical means – conjuring Berlin and Potsdam for desert nomads – was taken for real.

The story concludes by recording the establishment of a doppelgänger factory to replicate existing human presence and with the prediction that 'In the not too distant future, there will be whole cities made of light.'[7]

Friedlaender's remarkable story recapitulates, in certain respects, the experience of the retired magician Jean Eugène Robert-Houdin, who in 1856 was invited by the French military to perform magic before tribal assemblies in Algeria. This was driven by French anxiety about the political influence of the Marabouts, Muslim mystics whose authority seemed to be derived from their extra-mundane powers ('their ability to walk through hot coals, drink glass, lick hot metal ...'[8]). Napoleon III wanted to demonstrate that French magic was stronger than Marabout magic and to then reveal the 'trickery' in the hope that Marabout trickery would be simultaneously revealed. Accordingly, Robert-Houdin catches a bullet in his teeth and makes a 'young Moor' disappear in a cloth cone before revealing to the audience that his 'pretended miracles were only the result of skill, inspired and guided by an art called prestidigitation, in no way connected to sorcery'.[9] Later commentary suggests that we would be wise to be wary of Robert-Houdin's own assessment of his success: it appears that his audience found his efforts 'amusing' rather than compelling.[10] The film historian Murray Leeder usefully draws our attention to the broader target of Robert-Houdin's critique: Islam. Robert-Houdin repeatedly calls the Marabout 'false prophets' and identifies 'magical illusions masquerade[ing] as something it has no business being, religious experience, proof of the powers given to the Marabout by Allah'.[11]

Friedlaender and Kittler capture our attention with the audacious suggestion that cinema is merely a mirage by other means, the technological perfection of what is real but not true. But of course we shouldn't be surprised by this, for cinema was merely the culmination of centuries of phantasmagoric effort, from the catoptric cistula and magic lantern onwards, that has allowed men to compete with the wonders of nature. An index of this intimate relationship is the magician and subsequent film-maker George Méliès' purchase in 1888 of Robert-Houdin's Paris theatre, and the fact that his trick films were briefly known as 'Houdin films'.[12]

In the 1860s 'The Sphinx' caused a sensation on the stage of the London Polytechnic. This deployed mirrors to create the astonishing illusion of an Egyptian talking head in a box. Even greater excitement was stirred by 'Pepper's Ghost', 'a three-dimensional, specter-like figure [who] would appear on stage and walk through solid objects before fading away almost imperceptibly'.[13] Such wonders, frequently garbed in Orientalism, formed the infrastructure from which early cinematic pioneers such as George Méliès emerged. Méliès' infatuation with stage magic had its origins in his youthful obsession with the Egyptian Hall in Piccadilly. It was here that he may have encountered mirages, such as in a *Panorama of the Nile* in which after a 'fearful simoon, or sand storm . . . the Mirage unfolds its illusions to the eye'.[14] Appropriately, the Egyptian Hall had also earlier been the venue for Giovanni Belzoni's celebrated 1821 exhibition of ancient Egyptian treasures, which included a miniature replica of Seti's tomb replete with 'ibis-headed gods, snakes, demons without arms [and] mummies stretched on couches'.[15]

Mirages were also theatrically staged on behalf of 'science', and not just entertainment. The International Exposition in Paris in 1900 was the venue for Eugène Hénard's 'Palace of Illusions', which, following further technical refinement, would find a home at the Musée Grévin

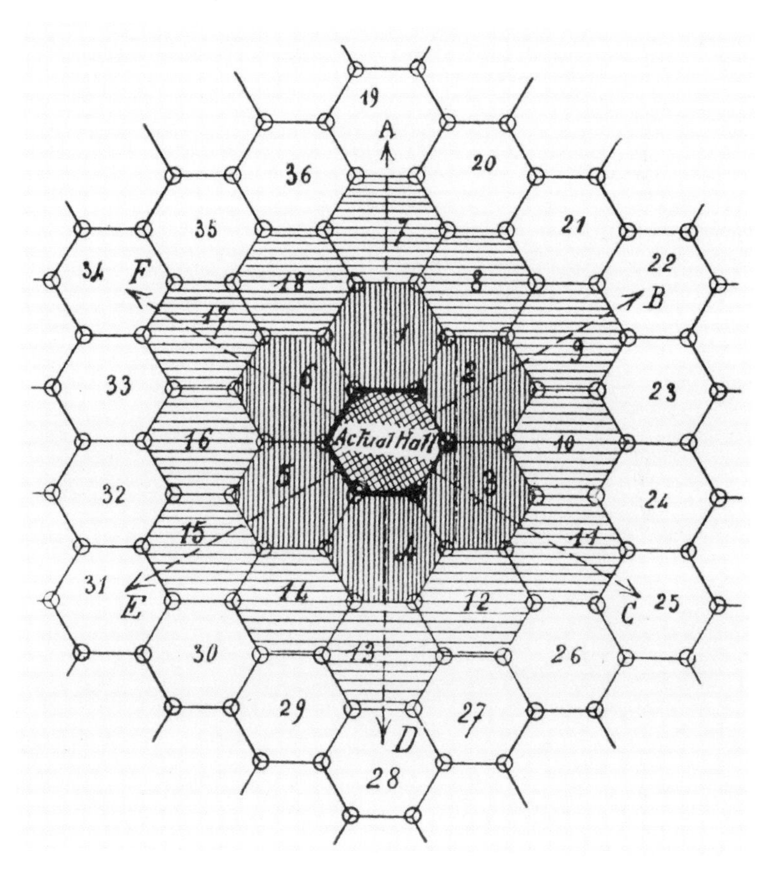


31 'A Diagram Showing How the Actual Hall is Reflected and Re-reflected Until Innumerable Identical Halls Seem to Exist', from the *Illustrated London News* (24 April 1909).

under the name 'Palace of Mirages'.[16] Hénard based his earlier effort on the Great Mosque of Cordoba, deploying mirrors to produce a 'sense of infinity'.[17] It is striking that while appearing, as one might expect, in the *Illustrated London News*, Hénard's device also featured in a lengthy report in *Scientific American* where it was presented as a version of the catoptric cistula described by Athanasius Kircher (illus. 31). The iteration of 1900, on the Champ de Mars, used mirrors framed by Moorish arches to produce a stunning Arabian palace and attracted two million paying customers. The later Palace of Mirages introduced further rotating mirrors, with suitable architectural motifs, that allowed for the creation of three environments: a Hindu temple (a benign version of what Rowlandson in his Elephanta phantasmagoria presented in negative form), a forest and an Arabian palace (illus. 32). The account in *Scientific American* provided first-hand testimony worthy of a travel account:

32 'The Hall of the Palace of Mirages as a Hindoo Temple', from the *Illustrated London News* (24 April 1909).

> We pass into a hall of apparently immense size, surrounded by mysterious galleries, which lose themselves in infinite depths. The roofs of these galleries are supported by massive columns in the Hindu style of architecture. The columns are surmounted by elephants' heads richly carved and studded with gems. Around the heads of the elephants great serpents are coiled . . . The silence of the hall is broken by a few musical chords. Suddenly the eyes of the elephants open. The stones which serve as their eyes and decorations glow in varied colored lights, and the great serpents shimmer like emeralds.

*Scientific American*'s correspondent was clearly enthused, perhaps to the extent of astonishment.[18] The excitement continues in the final catoptric shape-shifting extravaganza. Here the forest canopy disintegrates at the sound of a bell:

> the light gradually increases. In the faint illumination the spectator sees long avenues flanked by columns of luminous onyx. He wonders if he is at the court of a Khalif or in the Palace of Alladin. The dome shines like a sun . . . Mad arabesques seem reproduced infinitely. Everywhere stars shine as though in an endless firmament. Finally the entire dome blazes forth a brilliant glare of light.[19]

Linking Hénard's creation to Kircher's catoptric cistula provides a plausibly archaic manner of framing the Palace of Illusions and the Palace of Mirages. However, they also need to be positioned in the landscape of modern Paris, one that, as Elizabeth Carlson has observed, was increasingly dominated by plate glass and mirrored shop fronts. Just as the optical universe of the early nineteenth century was documented by publications called *Mirror*, *Kaleidoscope* and the like, so the flowering of mirage as metropolitan entertainment occurred within an environment – a 'city of mirrors', as Walter Benjamin called Paris – whose scopic potential was being radically extended.[20] For Benjamin, the proliferation of mirrors made the Parisian outside and inside endlessly transposable.[21] It radically reconfigured urban landscapes, jolting habitual public 'ways of seeing' through its 'illusions of odd, often contradictory spaces'.[22] Carlson adds empirical heft to this, noting that mirrors constituted between 3 and 4 per cent of new building costs in Paris in 1870, and that in 1900 Saint-Gobain (Hénard's sponsor) produced 90 per cent of the world's mirrors, making them (incredibly) France's biggest export.[23] In this context French 'ocularphobism' and Jacques Lacan's interest in the foundational importance of reflection seem hardly surprising.[24]

The Palace of Illusions, renamed the Palace of Mirages, speaks to the coherence of a linked set of ideas about the 'Orient' (in its Muslim and Hindu incarnations) and visual disturbance. The tenacity of these associations is suggested by its continuing presence in Paris. Relocated from the Champ de Mars to Musée Grévin after the close of the exhibition, it was rebuilt on that site in 2006 to mark the centenary of its new location. Its proud owners now advertise it as 'three dazzling displays' which 'transport the visitor to a bewitching Hindu temple, a hostile jungle and, finally a lavishly gilded Palace of the Thousand and One Nights'.[25] The forest has transformed into a threatening jungle, but otherwise everything remains the same.

The 45 different luminous effects facilitated by the complex arrangement of mirrors were controlled by a 'special keyboard' on which an 'electrician plays on a kind of switchboard piano'.[26] Photographs that accompanied both the *Illustrated London News* and *Scientific*

*American* coverage show an operator at the keyboard flanked by what looks like the console of an organ replete with stop controls. The electric 'switchboard piano' used to control lighting and the rotation of mirrors in this case directs our attention to the incipient synaesthesia in accounts of Hénard's device. Charles Leadbeater and Annie Besant's 1905 Theosophist classic *Thought Forms*, to which we will return later, took its prompt from Chladni's experiments with the graphic representation of sound. His 'sound plate' scattered sand over a brass plate, and when 'bowed' (with a violin bow) assumed abstract shapes which a later commentator termed 'vibration figures'.

33 Attrib. H. Jones, *The Pilgrim*, Richard Burton travelling to Mecca in disguise as 'Shaykh Abdullah', 1893, oil on canvas. Probably based on an earlier sketch by Edward Lear and Thomas Seddon made in Cairo in 1853. This image, used as a frontispiece in Burton's *Personal Narrative*, became the defining popular image of the contact zone between Europe and Arabia.

TWELVE

# The 'Mirage Medium of Fancy'

A decade before 'The Sphinx' so beguiled Londoners, Richard Francis Burton made his celebrated pilgrimage to Mecca in disguise. In an account of his travels in 1853 first published in 1855, Burton converted the Nabobism that hung as an accusation over Hastings into an enabling technique that allowed European vision to reach into formerly inaccessible regions. Any account of mirages must keep a special place for Richard Burton and indeed my title 'the waterless sea' is taken from his *Personal Narrative of a Pilgrimage to Al-Madinah and Meccah,* in which he gives the phrase as a translation from the Arabic *Bahr bila ma.*[1] Meaning 'desert', this phrase nevertheless magically captures the strange absence and presence that characterizes mirage. Burton describes leaving Al-Hijriyah and seeing in the early evening a mirage which 'completely deceived me' despite his being 'accustomed . . . to mirage'.[2] He further notes that 'beasts' are never deceived by mirages, which he attributes to their reliance on locating water through smell rather than vision. By 'beasts' Burton presumably means (exclusively) camels whose higher viewpoint would make them less susceptible to mirage effects. That this was so had been demonstrated by Giovanni Battista Belzoni. Born in Padua, he had come to England in 1803, where he made use of his great height (2 m, or 6 ft 7 in.) in a travelling circus where he was able to indulge his interest in phantasmagoria.

Initially engaged by the Pasha, Muhammad Ali, to demonstrate hydraulic irrigation equipment, he subsequently developed a career as an Egyptologist, removing the statue of the head of Rameses II (which remains in the British Museum), and starting the excavation of the tomb of Seti I. In his account of those activities published in 1820, he described the mirage as what the journal *The Kaleidoscope* described as one of the 'singular phenomena of Egypt'.[3] He narrates that commonplace paradox of mirage's seductive power ('I must confess that I have been deceived myself') before then describing an important experimental investigation into the effect of the angle of vision on the appearance of mirage:

> If the traveller stands elevated much above the mirage, the apparent water seems less united, and less deep; for, as the eyes look down upon it, there is not thickness enough in the vapour on the surface of the ground to conceal the earth from the sight. But if the traveller be on a level with the horizon of the mirage, he cannot see through it, so that it appears to him clear water. By putting my head first to the ground, and then mounting a camel, the height of which from the ground might have been about ten feet at the most, I found a great difference in the appearance of the mirage.[4]

Burton's experience of mirages is intimately entwined with his journey in disguise to Mecca (illus. 33). He was not the first European to infiltrate the ground zero of opacity, but he was certainly the most flamboyant and self-publicizing (his story, a later commentator noted, had been 'told by himself in three volumes of no mean dimension' and according to a writer in the *Athenaeum* was 'one of the dullest

34 *Mirage in the Desert*, *c.* 1885, chromolithograph. This wonderful realization of the 'water of the desert' was one of many illustrations in the famous German encyclopedia *Meyers Grosses Konversations-Lexikon*.

and worst written books of its kind').[5] However, like the *Arabian Nights*, Burton's narrative is peppered with visual illusions and effects. Arriving at Medina, Burton views the newly arrived Damascus caravan with its assortment of tents: 'The plain assumed the various shapes and colours of a kaleidoscope, and the eye was bewildered by the shifting of innumerable details.'[6] (illus. 34)

The two deceptions – desert mirage and Burton's own disguise – are triangulated in his narrative with the Kaaba in Mecca. There is a fascinating symmetry between Burton's own presence and that of mirage. Burton would famously assume the identity of 'Shaykh Abdullah', a wandering 'Darwaysh' (dervish), since 'No character in the Moslem world is so proper for disguise as that of the Darwaysh.'[7] The image of Burton in disguise was used as the frontispiece to his account of the journey and became a leitmotif of what we might think of as the interception of deception, by deception. The question of disguise was sufficiently contentious for Burton to address it at some length in the Preface to the third edition of the *Personal Narrative* in 1879. Burton focuses on a number of what he describes as 'truculent attacks' that demonstrate the ways in which reliability of vision, transparency and identity were clearly topics that aroused great passion. Burton discusses the 'foul blow' to the reputation of Johan Ludwig Burckhardt (himself a great eulogist of mirage) and the 'invidious remarks' directed at Ludovico di Varthema (who had disguised himself as a Mameluke), who had been condemned as 'renegade' for his 'deliberate and voluntary denial of what a man holds to be truth'.[8] Burton then engages in a fascinating onslaught by the Jesuit Arabist William Gifford Palgrave, who had travelled throughout the Middle East as a Christian in the 1860s. Palgrave thought that 'passing oneself off for a wandering Darweesh' was a 'very bad plan' and went on to critique the feigning of 'a religion which the adventurer himself does not believe, to perform with scrupulous exactitude, as of the highest and holiest import, practices which he inwardly ridicules, and which he intends on his return to hold up to the ridicule of others'.[9]

Palgrave, however, was no better than 'Satan preaching against sin', Burton declaimed, for he was himself the master of deceptive disguise (born a Protestant, of 'Jewish descent', a convert to Catholicism, reverted to Protestantism; English by birth but living under French protection; travelling in the garb of a 'native [Syrian] quack' and so on). If true, this perhaps explains Palgrave's own fierce preoccupation with the familiar trope of the duplicity of mirage. He writes, in ways that will now be familiar, of 'lakes of mirage *mocking the eye* with their clear and deceptive outline' and, later, of 'deceptive pools of the mirage'.[10]

Burton's defensiveness about his disguise may provide one way of situating the 'complete' deception that is mirage in his account. We have seen how, although 'accustomed as I have been to mirage', a vision in the vicinity of Al-Hijriyah 'completely deceived me'. On the horizon he sees 'fort-like masses of rock which I mistook for buildings', only discovering his mistake through the 'crunching sound of the camel's feet upon large curling flakes of nitrous salt overlying caked mud'.[11]

Burton's adoption of disguise and experience of mirage are all a consequence of his desire to enter Mecca and see the Kaaba, at the centre of Islam's holiest mosque, the point zero towards which *qibla* (the direction of prayer) is oriented, and the focus of the *salat* (prayers) during the *hajj* pilgrimage (illus. 35). The Kaaba is an origin point that bends space. Burton notes how it is only here that pilgrims can pray all around it in a circle. A black granite cube of about 12 m (40 ft) in height, there is something elemental about this primal object whose mystery is enhanced by the presence of the Black Stone, a silver-enclosed primal eye, set into its eastern corner. Dropped from the heavens by angels for the benefit of Adam and Eve, it might also be a meteorite. Burton tries unsuccessfully to get near enough to the stone to touch and kiss it but ecstatic pilgrims bar his way.

The Kaaba, the destination of all this dissembling and mystery, precipitates an extraordinary description when first sighted:

> There at last it lay, the bourn [destination or goal] of my long and weary Pilgrimage, realising the plans and hopes of many and many a year. The mirage medium of Fancy invested the huge catafalque and its gloomy pall with peculiar charms . . . the view was strange, unique, and how few have looked upon the celebrated shrine! . . . It was as if the poetical legends of the Arabs spoke truth.[12]

35 The Kaaba, Mecca. Detail from a popular offset lithograph, India, *c.* 2015. This is the edifice that provoked Richard Burton's memorable description of the 'mirage medium of Fancy'. Its decisive blackness fuses the sacred with the occluded, a seeming void punctuated by the ocular Black Stone.

In a characteristically pugnacious aside, Burton notes that

> the civilized poets of the Arab cities throw the charm of the Desert over their verse, by images borrowed from its scenery – the dromedary, the mirage, and the well – as naturally as certain of our songsters, confessedly haters of the country, babble of lowing kine, shady groves, spring showers, and purling rills.[13]

Burton clearly has in mind Gray's 'Elegy Written in a Country Churchyard' and his comments are an unlikely echo of the Episcopalian Joseph Packer's suggestion in the 1830s that Arabic poetry is 'deeply imbued with the characteristics of the climate – its burning sun – its wide extended deserts, where the eye perceives only the vault of heaven or the deceitful mirage'.[14]

Key figures in recent Arabic poetry, perhaps unavoidably, continue to have recourse to those tropes commented on by Burton. The Cairean poet Ibrahmin Naji (1893–1953) was known for the *Mirage Poems*, a response to unhappy love, which had turned life into 'a desert with its deceitful mirage'. This may appear clichéd, but as M. M. Badawi notes, 'the word *sarāb* (mirage) is a potent word in his poetic diction.'[15]

Alan Jones's illuminating consideration of the specific qualities and nuances of mirages, 'shimmering and gossamer', become even more interesting when we consider them in the context of the visible signs that often surround Jinns, those malevolent presences which find their way into English as 'genies'.[16] Jinns, as numerous illustrations testify, emerge from the space between atmosphere and fire. It is here that they seem to reconnect with mirage. Badawi and Haleem's *Dictionary of Qur'anic Usage* notes, 'and he created the jinn out of a smokeless flame of fire' (55:15, the fifteenth verse of Sura 55, 'The Beneficent'), recording *m-r-j* as indicating

36 *Buraq-ul-nabi*, on whom the Prophet ascended during his *mi'raj*. Chromolithograph published by S. S. Brijbasi, Delhi, *c.* 2000.

'smokeless flame, a bright tongue of fire', *mārij* as denoting 'a shooting out' of a 'flame of smokeless fire' and *marīj* as signifying 'confusing, and obscure'.[17] To add to the confusion that surrounds *m-r-j* we should note the coincidence of 'mirage' and that key episode in the Prophet's life, his *mi'raj*, night journey, or ascension from Mecca to Jerusalem on the steed Burak (illus. 36). Recall that the Arabic term for mirage is *sarāb* and that 'mirage' in English comes to us via French from Latin.[18] This etymology and the coincidence of *mirage* and *mi'raj* is clarified by Sir John Malcolm (who subsequently became Governor of Bombay) in his *Sketches of Persia*, published in 1827. He described his amusement at seeing, on the march from Abusheher, 'that singular vapour (capable of making a man at a distance of a mile and a half "as tall as a date tree") called by the French Mirage and by the Arabs and Persians Sirab'.[19]

*Mi'raj* is not simply 'mirage': on this we should be clear. And yet it is not wholly disconnected. Muslims drew parallels between

Jacob's ladder vision and Muhammad's *mi'raj*, and the Arabic text *Kitab al-mi'raj* (the Book of Mi'raj) was translated in the thirteenth century as *Libro della scala* or *Livre de l'eschelle* (Book of the Ladder).[20] *Book of the Ladder* describes the Prophet's journey through the various levels of heaven, including

> a series of gardens abundant in wonderful pleasures: beautiful women . . . walls and houses constructed of brightly colored jewels flashing with light, which delight the eye; and rivers flowing with delicious liquids including milk, honey, and wine.[21]

And, echoing imagery with which we will already be familiar, vision is overwhelmed by light, visions of paradises and gardens made 'wholly of light [and] cities and castles [made] entirely of light'.[22]

One does not need to fully decode the iconography of William Blake's visionary painting of *Jacob's Dream* from 1805 to be persuaded that the vision it records might be persuasively inserted into a spectrum that embraces dreams, crepuscular rays and mirages. Dreamt, originally, in the deserts of West Asia, it comes from an environment saturated with the smoky *sfumato* of mirage. 'Smoke' of course appears as a metaphor in several accounts of mirages. Alexander Burnes, journeying through the Rann of Kutch in Gujarat in the 1830s, eulogized salt marsh as the perfect habitat for atmospheric refraction:

> Nowhere is that singular phenomenon, the *mirage or surab* of the desert, seen with greater advantage than in the Run. The natives aptly term it smoke; the smallest shrubs at a distance assume the appearance of forests; and on a nearer approach, sometimes that of ships in full sail.[23]

THIRTEEN

# Mirage and Oriental Despotism

The potency of mirage derives from an apophatic insistence on what it is not. Originally associated with 'negative theology', apophasis, like the rhetorical figure of litotes (in which 'an affirmative is expressed by the negative of the contrary' as the *SOED* puts it), provides a way of understanding the appeal of the phrase 'the waterless sea' and also of the mirage as deception or mockery. Mirages (water, palace and so on) are seductive and alluring but always underwritten by a falsity. Their positivity rests on the foundation of being other than they appear. Litotes involves what Jonathan Lamb terms a 'ghostly specificity',[1] often entailing detailed descriptions of landscapes of absence, topographies of things that are not there.

This apophatic quality extends, as we have already begun to see, to its 'un-Christianity', which in European travel narratives can be both negative and, less commonly, positive. Mirage as the positive antithesis of a disenchanted Christianity is memorably presented in Henry Miller's *The Colossus of Maroussi* (1942). In this beautifully evocative description of Greece during the onset of war, he describes Athens as floating 'in a constantly changing light, beat[ing] with a chromatic rhythm. One is impelled to keep walking, to move on towards the mirage which is ever retreating.'[2] He then announces that on the Sacred Way, walking from Daphni to the sea, he was 'on the point of madness several times' surrounded by 'trees flooded with light, intoxicated, coryphantic trees which must have been planted by the gods in moments of drunken exaltation'.[3] The stones and trees are remnants of a paganism suffused by a holy light which Miller describes in exquisite prose:

> No analysis can go on in this light . . . The rocks themselves are quite mad: they have been lying for centuries exposed to this divine illumination: they lie very still and quiet . . . but they are mad . . . and to touch them is to risk losing one's grip on everything which once seemed firm, solid and unshakeable. One must glide through this gully with extreme caution, naked, alone, and devoid of all Christian humbug.[4]

Miller provides the Utopian version of Camus' dystopian vertical fiery light in *L'Étranger* (The Stranger), published in the same year, 1942.

The metaphor of the mirage and this fantastical description of ocular lunacy are for Miller symptoms of an alluring ancient Greek paganism, which through a backward-looking exoticism offers the erasure of two millennia of Christianity. But in very many accounts, the mirage functions as a route to something more specific than that which is not Christian. Mirages, as I have suggested, are frequently explicitly associated with Islam, and with a form of politics diametrically opposed to what Tocqueville eulogized as 'spectatorial democracy'. Mirage's optical trickery, its 'mockery', comes to exemplify a subaltern opacity that appears to be the opposite of Euro-American acuity. Mirages serve to conjure what became known as 'Oriental despotism'.

The creation of Oriental despotism as the very opposite of Tocquevillean spectatorial democracy was able to draw on a huge repertoire of imaginative resources. Central to this was the *Arabian Nights*'s preoccupation with visual deception (through dreams, illusions and ocular apparatus) served as a vehicle for the worlds which the *Nights*'s narrative could invent and describe. This narrative tradition would come to fascinate and be endlessly invoked in European narrative and performance traditions, ranging from theatrical phantasmagoria and stage magic to novels, travel literature and film.[5]

The *Arabian Nights*, also known as the *Thousand and One Nights* (of which Burton produced his own highly personal translation), was itself a major vehicle of what we might term the enchantment of illusion. The stories' framing narratives are one symptom of this but the narratives are also littered with optical devices and uncertainties about the relationship between the visible and the true. Burton described his 1885 translation as a 'natural outcome' of his pilgrimage to Medina and Mecca and in his quite remarkable translator's preface noted how it was 'Impossible even to open the pages without a vision shifting into view; without drawing a picture from the pinacothek [picture gallery] of the brain.'[6] Burton asks us to picture him narrating the tales to 'Shaykhs' and 'white-beards' as well as women and children who are transfixed as they imagine the 'most fantastic flights of fancy, the wildest improbabilities, the most impossible of impossibilities [which] appear to them utterly natural, mere matters of everyday occurrence'.[7]

The story that Burton translates as 'The City of Brass' conjures a phantom city of the dead whose allegorical moral resonates with examples that we have already encountered. Although located in the Maghreb (specifically Mauritania), it echoes the Gandharva cities and might be evidence of the Indian lineage of much of the *Arabian Nights*'s narrative. A glittering city looms through the desert. Its 'lofty columns and porticos . . . inlaid with gold and silver and precious stone' seem to testify to opulence and success. But of course the city is deserted, filled only with the dead, and is transformed into a monument to 'greatness fallen into dust and clay'. The 'warnings' that the vision offers are further reinforced (pomps, vanities and sundry falsehoods are excoriated) before a whole catalogue of similes is offered: 'It is like unto the dreams of the dreamer and the sleep visions of the sleeper, or as the mirage of the desert, which the thirsty take for water; and Satan maketh it fair for men even unto death.'[8] Burton's translation tumbles together what we might expect to be quite distinct: the deception of the sense by inferior mirages, the divine revelation of dreams, and the profane trick of Satan. An edition of the *Arabian Nights* from 1885, edited by the Indian nationalist Romesh Chandra Dutt, made the miragistic dimension of the city quite explicit, recalling Hari Chand's city as described by James Tod, adding that Fata Morgana, Saint Borodino [Sant Brandon], Cape Fly-away and the *Flying Dutchman* hardly needed to be mentioned.[9]

A popular selection of *Arabian Nights* tales from 1909, illustrated by Maxfield Parrish, perfectly visualizes the City of Brass as a looming Fata Morgana (illus. 37). The Maghreb is incarnated as a slice of mountainous North America with three Arab horsemen looking almost like Native Americans as Edward Curtis would have imagined them. Striking diagonals define a dark foreground and a distant sunlit mountain from which emerges the spectacular rectilinear form of the City of Brass.[10]

Montesquieu's mid-eighteenth-century exercise in what Ros Ballaster has termed 'reverse ethnography', *Persian Letters*, presents a gentle French view of the virtues of European rationalism and a critique of Ottoman rule which would achieve a new virulence in twentieth-century preoccupations with the 'sick man of Europe'.[11] However, Montesquieu provides much of the basic infrastructure of this critique when Usbeck (purportedly the Persian author of the letters) writes to his friend Rustan in Ispahan with his thoughts on the 'weakness of the

37 Maxfield Parrish, 'City of Brass', 1905, included as an illustration in Kate Douglas Wiggins and Nova A. Smith, eds, *The Arabian Nights: Their Best-known Tales* (1909).

Ottoman Empire'. He has arrived in Smyrna in 1711 and has spent 35 days crossing Turkey. In his letter, he bemoans Ottoman corruption: the Pashas have bought their positions, the army is unruly, private property is not guaranteed, and the citizens of the empire 'suffer innumerable abuses of power'.[12] The Ottoman Empire is denounced as 'a diseased body, preserved not by gentle and moderate treatment but by violent remedies which ceaselessly fatigue and undermine it'.[13]

The Ottomans are contrasted with the intellectual duels in evidence in Paris's coffee-houses 'where the coffee is prepared in such a way that it sharpens the wits of those who drink it . . . there is nobody [who] does not think that he is four times cleverer than when he went in.'[14] This is an early incarnation of what Jürgen Habermas would later term the 'critical rationality' that characterized an emergent public sphere. Usbek is won over by 'mild' government, the kind that 'attains its purpose with the least trouble' and decides in favour of 'reason' rather than 'harshness'.[15] Religious intolerance and proselytization is pronounced 'dizzy madness, the spread of which can only be regarded as the eclipse of human reason'.[16] Later still, Usbek finds himself persuaded by 'natural science', which 'five or six truths' has filled with 'miracles'. [17]

In addition to those critiques (such as Montesquieu's), which we might label 'civilizational', there are, unsurprisingly, a large number of theological attacks which mobilize the 'light' of a preferred religion against the 'darkness' or iniquity of a competitor creed. J.W.H. Stobart's *Islam and Its Founder* of 1876 is typical of nineteenth-century Christian missionary discourse but one that is more worthy of attention than the average sectarian assault because of its explicit focus on questions of what we might think of as 'civil society'. Stobart was principal of the famed La Martinière College in Lucknow and much of his text is detailed and empathetic. But his analysis then reaches a crescendo of Christian fervour, which is striking for its foregrounding of 'life, liberty, and social and political well-being', a preoccupation that is doubly ironic in the context of colonized India:

> The Koran claims to be a continuation of the earlier messages of Heaven, and to supplement and develop the teaching of the Law and the Gospel. Assuming such to be the case, we may fairly look to it to afford us clearer views of the Divine will and attributes of life and death, of the provision made for man's spiritual and temporal difficulties; and in it we should find the way made more plain for securing to all mankind their inherent rights of life, liberty, and social and political well-being. Instead of this, darkness and retrogression are engraved on every page of the 'Preserved Book,' God's universal fatherhood is ignored, and in place of the finished sacrifice, the sinner is bid to plunge into the dark future, trusting in his own righteousness.[18]

For Stobart, the 'loving dictates of the Gospel' are 'light' whereas the 'denunciations of the prophet' are 'darkness'. The occlusion of Islam stands opposed to the transparency of 'social and political well-being'. This is the Christian missionary version of Tocqueville's 'spectatorial politics'.

The nakedly political dimension of mirage as a metaphor for Islam was most starkly revealed in an anonymous story, 'A Romance of the Mirage', published in *The Cornhill Magazine* in 1883. Although a 'romance', it is the elaboration of information given to the author by an official of the Telegraph Service 'as we steamed one morning across the blue bay of Suez' as (inevitably) 'a slight mirage lay beneath the glowing hills on the desert edge'. The Telegraph Service official had previously worked at a remote station – Um el Jemal – on the Red Sea, helped by two 'native clerks' and two servants. This remote station was 'the home of mirage', and mirages made the desert 'lively': 'Fishing craft sailed in pellucid rivers; sometimes a great merchant ship or a man-o'-war appeared; villages stood out distinctly, camels and caravans stalked along, men prayed and marched.'[19]

Most of these mirages were ever-changing, but one remained the same, appearing day after day. This took the form of an ancient castellated structure with a huge gateway: 'It did not glimmer into view, nor flicker in vanishing, but burst on the eye complete.'[20] One of the clerks, named Zohrab, took a great interest in this mirage, closely

38 Zohrab beholding the distant Wahabi mirage, detail of an illustration by R. C. Woodville, from 'A Romance of the Mirage', a story published in *The Cornhill Magazine* in 1883.

observing it and its white-robed visitors whose 'frequent praying and preaching, told a political secret' (illus. 38). The mirage was the spectral mansion of the Wahabis and 'therefore a home of treason and rebellion'. Wahabis, or Salafis, were followers of Muhammad ibn Abd al-Wahhab (1703–1792), whose advocacy of an austere Sunni Islam would find common cause with Muhammad bin Saud, whose descendants would establish the kingdom of Saudi Arabia in 1932.

Zohrab was fascinated by what he saw. Although a Christian, he was a 'fanatic patriot' and 'hated the Wahabi schismatics almost as bitterly as they hate his own creed', but reluctantly concluded that 'the supreme foe, the Turk, will only be expelled by the aid of these blood-thirsty desperadoes.'[21] The grand conspiracy in the mirage started to haunt him. Zohrab invented a persona for the sheikh whose black *burnous* (long woollen cloak) and highly coloured clothes were difficult to reconcile with his Wahabi identity. Zohrab granted him a daughter, Ferideh, whom Zohrab imagined he would marry after the declaration of Arab independence in Damascus and the beheading of the 'false Khalif' (the Ottoman Sultan).

The mirage and the stories he had invented about its inhabitants completely obsessed Zohrab. He drew plans of the castle and gave them to sailors on the supply boat from Suf asking them for more information. After many months, there was success: a Bedouin, passing through Suf, had immediately recognized the plan. It was El Husn, 'the fortress-palace of Sheikh Abou 'l Nasr, lying at a distance of four days' journey across the desert from Suf. Abou 'l Nasr was a former Wahabi who had grown not only rich on his spoils from Mecca and Medina, but wise, and a great expert in magic. Being told that he must relocate to a new telegraph station in Egypt, Zohrab takes the supply boat to Suf and rides towards the mirage. He is captured by Wahabis who then take him to El Husn which, as he approaches it, finally reveals itself to be the place known through the mirage. There follows imprisonment, battles in Suf and marriage to another of the Sheikh's daughters, who converts to Christianity in India, where the couple, and the story, find a happy ending.

# FOURTEEN

# Keeping Mecca and Medina Invisible

It would be a mistake to enthrone Enlightenment optimism, or missionary fervour, as determinants of the European world-view that generated so many visual and literary discourses about mirages. They were one aspect of a much more complicated story. Burton's enthusiasm for the 'mirage medium of Fancy' expresses that very common positive Orientalism that Ronald Inden labelled 'the loyal opposition', an embracing of the 'Orient' as a repository of an enchantment that progress and history had destroyed in the West.[1] Arthur Rowan, a squadron leader and spy whose career took him to India and Egypt in the early twentieth century, provided a memorable example of this in his description of an oasis near Cairo: 'I felt that in such a place – in this expanse of boundless wilderness – we could not fail to recover for ourselves something of what we had unavoidably lost: that here we must surely "find ourselves anew, and communicate once again with the old, forgotten things of lost wisdoms".'[2]

The nascent 'critical rationality' that Montesquieu celebrates promised a transparency that for other commentators offered a threatening disenchantment. Augustus Ralli opens his entertaining early twentieth-century overview of European travellers to Mecca by noting that when he once told a friend that 'a visit to Mecca would be of incomparable interest' the friend had advised not to visit during the

Pilgrimage (Hajj). When Ralli replied that 'no Christian might look on the holy city of Islam and live, [the friend] asked in astonishment if the prohibition extended even to Cook's tourists!'[3]

Mecca, and hence Islam, embodied the antithesis of the 'spectatorial mode of politics' advanced so clearly by Tocqueville. If American democracy offered transparency and visibility, Mecca was a zone of occlusion policed by absolute violence. And yet Ralli ends his compendium with the possible triumph of the former over the latter. His concluding chapter concerns the Hejaz railway, the last section of the Baghdad railway that would link Constantinople to Mecca (linking 'the Sultan of Turkey's temporal and spiritual capitals').[4] By August 1908, just before the publication of Ralli's book, the line had reached Medina. The Prophet's tomb had been 'illuminated for the occasion with electric light' and a message from *The Times*'s special correspondent was telegraphed in English from Medina. That correspondent speculated that it was only a matter of time until 'the snort of the locomotive will . . . be heard within the precincts of the Kaabah.'

For Ralli, this was not the cause for wholehearted celebration, for 'with the linking of the Moslem holy cities to the world's great life-centres, something of their mystery will be absorbed in universal circulation.'[5] The 'wild journey[s]' of the past will no longer be necessary. Tourists will steam to Jeddah and then take the branch line to Mecca. The traveller will see 'framed in the window of his saloon carriage, a picture of the terrifying Arabian landscape'.[6] Alongside this disappointment, Ralli also recognizes the triumph of Tocquevillian transparency. He reports an orator at the opening ceremonies claiming that '"The Prophet" . . . did not permit the railway to reach Medina until the Caliph had granted a constitution to his people' and a writer in a contemporary journal proclaiming that 'Now that the electric light burns over the Tomb of the Prophet, we may hope some day to see with our own eyes the sacred cities of the Moslems.'[7] Ralli's ambivalent suggestion seems to be that electricity and train travel will dissolve the 'mirage medium of Fancy' that drove most of the adventurers who feature so heroically in his account.

# FIFTEEN

# Inside Abdul Hamid II's Head

This part of the history of mirages cannot be reduced entirely to an opposition between Christianity and Islam: the imaginative role of the Ottoman Empire as the 'sick man of Europe' is also important. We have already heard a trenchant evaluation of the Ottomans from Montesquieu's Persian travellers. Richard Davey's *The Sultan and His Subjects* (1907) provides a strong early twentieth-century view of the 'sick man'. Recalling conversations with his old friend 'Dick Burton' (whose *Arabian Nights* translation he notes, as an aside, 'is certainly not reading to be recommended to the young ladies of a convent school'), he remembers his admonition that the 'people of the East' need to be seen 'by their own brilliant light, and not by the dim glimmer of a London fog'.[1]

For Davey, Istanbul was a mirage waiting to fall apart. The traveller might imagine himself about to 'set foot in a fairy city' but upon landing, 'great is his disillusion!', for the city supposedly guarded by God will be found to be infested with dogs living on filthy dunghills.[2] This was one symptom of what he called the death of Islam, a religion that he believed was incapable of progress and in which the orthodox Mohammedan disdains those who 'venture to lift even the corners of the veil which Allah, as he believes, has purposely drawn'. Tocqueville's political theory of transparency shines

39 W. D. Downey, 'A Young Abdul Hamid II'. Reproduced as the frontispiece in Davy's *The Sultan and His Subjects* (1907), the image was originally sold as a cabinet card.

through this caricature of the Ottoman lifeworld; one characterized, writes Davey, by a 'dreamy existence'.[3]

The frontispiece to Davey's *The Sultan and His Subjects* featured a photograph of a youthful, and wide-awake, Abdul Hamid II taken by the London photographer W. D. Downey (illus. 39). The dreamy existence of the sick man of Europe was given a remarkable miragistic form in a series of early twentieth-century postcards depicting the elderly and by now extremely reclusive Sultan Abdul Hamid II anamorphosistically with a face composed of voluptuous entwined naked females and a fez-enclosed head of empty oasis-ian night-time dreams (illus. 40). These popular images (of which there are many more iterations, including Japanese ones) may well have been inspired by a prototype anamorphosistic image labelled 'Fata Morgana' (illus. 41). The image here might plausibly be read as a sailor being charmed by Morgan le Fay, the sister of King Arthur, who liked to lure seafarers into her submarine palace. As Marina Warner notes, an Italian variant has Morgan le Fay grant her lover eternal life; in order to cope with his boredom, he is forced to conjure fantastic spectacles, which is perhaps what we see playing over his face.[4] Whatever the precise antecedents of the popular Fata Morgana image, it seems clear that the dreamy Sultan Abdul Hamid II comes out of this tradition of representing an individual under the influence of Morgan le Fay. Fata Morganas are clearly the iconographic prototype for an extraordinary image of political corruption and decay.

The Ottoman Empire, for reasons that affirm the central hypothesis of this book, became the preferred site of mirage experiences. It combined both Islam and what was seen to be an increasing occlusion with ongoing military conflict with much of Europe. There is a long history to this. One of the very earliest of mirage descriptions, in Antonius de Ferraris's *Liber de situ Iapygiae* (Book of the Situation

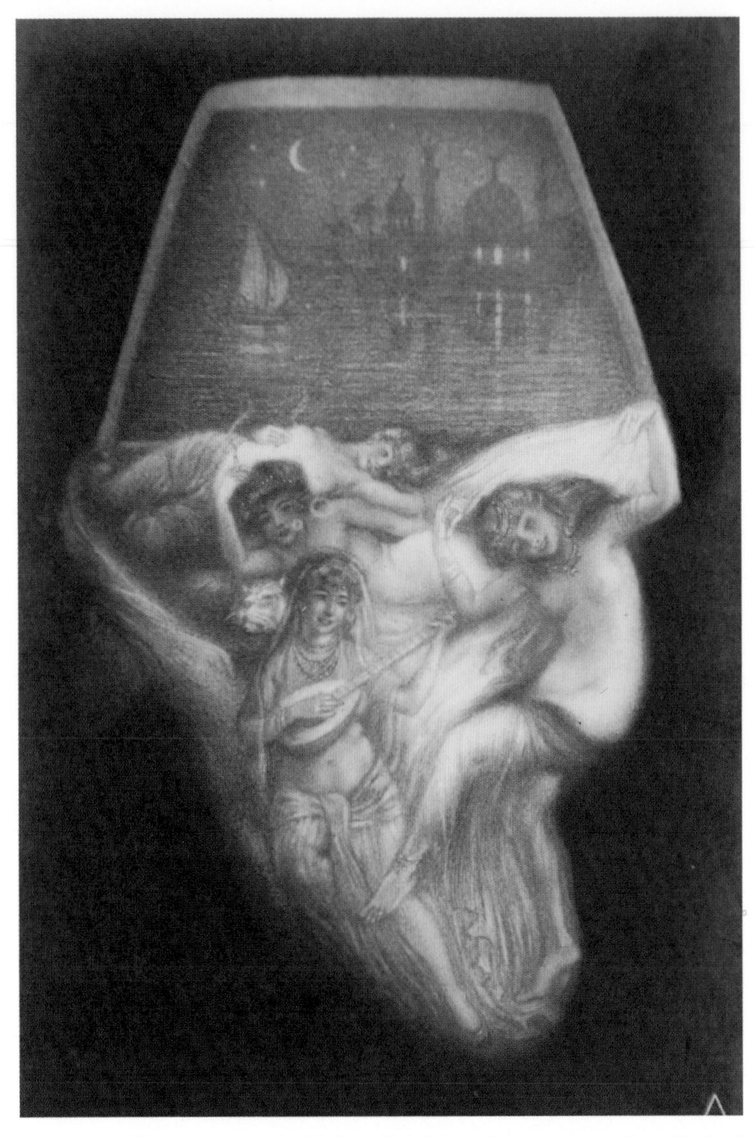

40 Anamorphosistic postcard of Sultan Abdul Hamid II, *c.* 1905, produced by Paul Fink, Berlin. The 34th Ottoman Sultan's face conjures the pleasures of the harem while his comparatively empty brain conjures a mirage-like vision. Abdul Hamid II (1842–1917), who oversaw the decline of the Ottoman Empire, suffered from increasing paranoia, resulting in his loss of the throne in 1909.

41 *Fata Morgana*, postcard, *c.* 1900. This early anamorphosistic image depicts a (presumably) Sicilian sailor on whose face flicker incarnations of Morgan le Fay, offering her lovers eternal life.

in Apulia) of 1558, following a remarkably insightful discussion of refraction, folds the phenomenon into a very current political anxiety:

> And as these figures are of mists, they give likenesses of ships and sails, where there is no fleet. These apparitions deceive not only the inexperienced. It is not long since the whole coast, from Hydrunto [Otranto] to Monte Gargano, at one and the same hour before sunrise, saw a fleet sailing from the east. It was thought to have been that of the Turks, and before that specter or delusion was revealed by the lightening dawn, various letters were composed here and there and messengers were sent concerning the approach of this imposing fleet.

As A. T. Young notes in his remarkable bibliography of mirage literature, Ferraris (1444–1517) had assisted in the liberation of Otranto on the Italian Salento peninsula following its capture by the Turks. In 1480 Mehmet the Conqueror had invaded Otranto with the aim of conquering Rome, and had (so Christian myth has it) beheaded eight hundred locals who had refused to convert to Islam during the thirteen months of Ottoman rule. Then again, in 1537, the Ottoman admiral Barbarossa briefly recaptured the city.[5]

Three and half centuries later, Ottomans, facing the collapse of their empire, were still entangled in mirages, as evidenced by Martin Swayne's immensely readable *In Mesopotamia*. Swayne was a medical doctor employed during the First World War Mesopotamian campaign and presents mirages as a perplexing problem for science. He establishes this first and foremost as a problem of rationality through the appearance of a patient, a New Zealand wireless operator, who 'had found something in the desert that puzzled the science of his mind'.[6]

Much of what is initially described in the 'white undulating expanse, burning hot' is benign enough. Swayne notes that a mirage has 'a pleasant side when it paints water and palms' or produces other diverting effects: 'A man standing a short way off may assume gigantic proportions, or look like a dwarf. A motor car near by would seem to lose its solidity and dissolve into a few filmy lines.'

But it quickly assumes the taint of war and a challenge to rationalism: 'An Arab might lie in the open and no one would see him. A post might look like a horseman at full gallop. It was a country of topsy-turveydom as regards the subjective estimate of the eyes.' This is mirage's 'sinister side when it clothes the most innocent features of the landscape in images of dread'. The wireless operator thought he had grasped the science of refraction (what Swayne calls the 'physical considerations') that was responsible for these illusions. But what he struggled with was the collectivized nature of subjective experience: how was it that 'several men would experience exactly the same illusion: why should a post simultaneously appear as an Arab on horseback or an Arab crawling stealthily on the ground to half a dozen men?' This for Swayne is the nub of the problem: how to remain sane and enact the rational protocols of modern warfare: 'Our gunners found this a continual difficulty at the front, for hostile Arabs, knowing the mirage areas, would get into them making ranging impossible.'[7] A later Smithsonian Institution report by the Arizonan meteorologist James Gordon (who was unable to provide sources) enrolled a mirage in the war *against* Ottoman forces: 'It seems that the British and Egyptian troops were fighting the Turks. The Turks were winning when they got word of a coming flank attack. The Turks withdrew. But there was no flank attack. It was a mirage.'[8]

Abdul Hamid II's head full of dreams is not only an Orientalist calumny, for it also echoes, albeit in a highly pejorative manner, a key part of Muslim intellectual commentary on the role of dream visions. Ibn Khaldun, in what is widely recognized as the first work of world history, *The Muqaddimah* (written in 1377), has many remarkable things to say about dreams, not least the manner in which they are almost (I stress *almost*) the result of an internal physiological process of refraction. Dream visions, he notes, are approved by Muhammad as the 46th part of prophecy, and are to be valued because they occur within sleep when the 'hindrances' of 'external senses' are lifted.[9] Ibn Khaldun describes 'internal refraction' in the following manner: the senses are distributed by a vapour which spreads from the left cavity of the heart (here Ibn Khaldun draws on Galen), the finest part of which goes to the brain. This sense-spirit is warm and is 'tempered by the coldness of (the brain)'.[10] What we might think of as a heat-gradient is amplified by the cold that 'covers the body during the night' and forces the warm sense-spirit to seek refuge in the interior of the body, causing sleep. Here Ibn Khaldun and Orientalist stereotype are revealed as inverted mirror images of each other. For the Orientalist, sleep signifies torpor; for Ibn Khaldun, sleep allows the triumph of inward powers over external sense-spirits, cleansing a path through which 'clear dreams' can come directly from God.[11]

The conjunction of dreams and mirages in relation to the Ottoman Empire can also be approached through another, de-historicist route: Ismail Kadare's remarkable novel *The Palace of Dreams*, first published as *Nepunesi i pallatit te endrrave* in 1981 in Albania, where it was promptly banned. Kadare's book is clearly a satire about totalitarianism (the reason for its ban), but its explicit subject is a bureaucratic institution, the Tabir Sarrail, on the periphery of the Ottoman Empire. Though 'one of the pillars of the state', dedicated to analysing the dreams of all its citizens, Kadare stresses that the Tabir Sarrail is 'an institution totally closed to the outside world' and narrates the bewilderment of his Christian Albanian protagonist, Mark-Alem, on being offered a job analysing the dreams of all citizens ('a vast enterprise, beside which the oracles of Delphi and the predictions of all the hordes of prophets and magicians in the past are derisory') in order to identify those in which 'Allah looses a forewarning'.[12]

The dreams requiring such close monitoring seem to operate (like mirages) according to a meteorological protocol ('the ebb and flow of dreams . . . varied according to the time of year, rainfall, temperature, atmospheric pressure and humidity'). The dreams conform to the perceptual regime of mirages, for they are real but not necessarily true: 'visions which to the ordinary eye seemed like meaningless mental doodling' revealing phenomena which in a 'logical world . . . would be absurd'.[13] And Mark-Alem encounters sights such as the arrival of the dream-carrying horses and wagons late at night, whose 'lurid glow of . . . lights . . . produced an almost supernatural spectacle'.[14]

In a still different idiom, a recent anthropological study of dreams and their interpretation in Egypt has provided wonderful insights into the knowledge that dreams offer and the contentious nature of their diagnosis. Amira Mittermaier documents the popularity of dream interpretation with manuals available at most Cairo bookstalls and television shows featuring dream interpreters. One TV show, *Ru'a*, featuring a sheik, a Sufi and a psychologist, supposedly interpreted a dream in which the moon breastfed a boy as foretelling the birth of the *Mahdi* (the prophesied redeemer). This led to the termination of the programme following pressure from orthodox religious spokesmen (including the Muslim Brotherhood and sundry Salafis) and liberal intellectuals who objected to what they saw as superstition.[15]

SIXTEEN

# Mirage *Pharmakon*: Wild and Domestic

Mirages frequently serve as a symptom of something gone wrong, something occluded that ought to be transparent, but they are often also symptoms that fascinate and bewitch. As Kant concluded, writing about the Beautiful and Sublime, 'Whoever loves and believes the fantastic is a *visionary*,' and mirages served for many as the ground zero of vision and visionary possibility where insight might spring from blindness.[1] This mirrors the dynamic that Friedrich Kittler hints at in *Optical Media* in his description of the Reformation's attempt to abolish the visual in favour of language. The Reformation, he writes,

> abolished or literally blackened medieval church rituals, with all of their visual glitter, and replaced them with the monochromatic, namely black-and-white mystery of printed letters. *Sola sciptura, sola fidei* – solely from writing and solely from belief.[2]

It was against this background that the Fata Morgana enthusiast Athanasius Kircher and other Jesuits embraced the magic lantern to restore some of that glitter and the mysteries of sense-experience. Recall that he had organized public projections of Fata Morganas, complete with 'moving warriors and devils upon a screen'.[3] 'Fever and delirium, books and holy images' combined with endless exercises to image the experience of hell to produce 'psychedelic visions' in the cause of a new holy war. The success of the Jesuitical Counter-reformation, Kittler continues, was built not only on the millions of deaths in the Thirty Years War but through what he calls 'bright' power, a 'new kind of imaging'.[4] Through Athanasius Kircher's magic lanterns, the Jesuits established 'a new kind of image worship, which, like the new hallucinatory readings [of Loyola, were] aimed at transferring the psychedelic effect of the spiritual exercises to the outside'.[5]

Writing in the 1870s, Georg Hartwig, the voracious documenter of all things natural and saleable, recorded the *pharmakon* quality of *serab* (mirage) and *bacher el afrid*.[6] The latter is the 'sea of the devil'. The former is akin to the 'works of the unbelievers', according to the Koran, which predicts that 'the thirsty traveller mistakes them for water, until he approaches them and finds that it is for nought'.[7] This points to a paradox in which the mirage acts as a two-sided mirror, as though in complex catoptric cistula, in which two parties on either side see the inversion of what they themselves are. So Burton's obsession with the Kaaba as a black void, a delusional Fata Morgana, is possible, perhaps, only because he is in disguise. He has to pursue trickery and illusion in order to reveal the other's delusional claims in a kind of pathological, oppositional version of Benjamin's mimetic faculty.[8] Benjamin had argued that the human ability to see resemblance was 'a rudiment of the powerful compulsion ... to become and behave like something else'.[9]

As it turns out, this tension at the heart of mimesis between alterity and likeness is predicted in the empirical experience of certain atmospheric anomalies, in particular the kind that would become most famous as the 'Spectre of Brocken'. These were associated with misty mountains (such as the Brocken, the highest peak in northern Germany), where at high altitude, mist 'arising from the valleys sometimes show[ed] the spectator his own reflected image on a gigantic scale'. Empirical observation demonstrated that these monstrous yeti-like figures took after their spectators. In the case of the traveller Dr Haue in 1797, he encountered a figure of monstrous size and then noticed the spectre mimicking him: 'he made a stooping gesture and the spectre stooped; in short all his gestures were repeated by his own magnified and shadowy image' (illus. 42).[10]

Hartwig points out that there have been numerous (less spectacular) incarnations of the 'Brocken effect', even in the unlikely environs of New Radnor (on the Powys/Hereford border), where in 1851 a Miss E. ascended the hills above the valley, picked flowers and waved her handkerchief, before descending to a point where she could see both the ascent and her friends below. She and her friends saw another figure identical to her own: 'The dress and flowers were precisely similar to her own and the colours were so vivid that she could even see her face. The effects were the same as if she were before a looking glass.'[11]

The 'mirror' potential of mirage (a strange anti-mirageism which eliminated refraction) demonstrated its political potentiality during the Siege of Paris (1870–71), when one of the many balloons that, alongside carrier pigeons, were used to ferry miniature photographic communiqués found itself reflected in a halo. Near-contemporary images suggest that this validated (through 'nature') the political resistance of the French nation in the face of Bismarckian bombardment. Mirages were overwhelmingly unruly, *unheimlich*, but some of them were tamed in the cause of nostalgia or of political affirmation.

Mirage and atmospheric effects, as we have seen, are most likely to manifest as distortion, like an ancient piece of hand-blown glass whose differential thicknesses produce refractive effects. But experiences of the Brocken kind suggest an uncanny mirroring of the beholder. The first idiom suggests fragmentation, and we have seen that elaborate Fata

42 'The Spectre of Brocken', from Camille Flammarion, *L'Atmosphère: Description des grands phénomènes de la nature* (The Atmosphere: Description of the Grand Phenomena of Nature, 1872).

Morgana distortions can be enormously complex. The metaphor of the mirrors suggests replication with stability and fidelity. However, mirage experiences can usefully be approached under other kinds of rubric. Consider 'domestic' and 'wild' mirages, for instance, and the parallel duality of 'territorialization' and 'de-territorialization'.

Territorialization involves a taming of the mirage, a turning of the savage dream, which Barthes once noted made him *foreign*, something both uncanny and familiar.[12] Perhaps the clearest example of this is provided by Alexander Kinglake in *Eothen*, a narrative of a journey to the Near East in 1834, in which he famously narrates an example of what has subsequently been rationalized as an 'acoustic mirage'.[13] It follows on from an account of a salt deposit mirage and is of particular relevance for the manner in which it invokes one location and identity in the place of another. Kinglake hears church bells, which he interprets as calls to Christian prayer in a location of placelessness. He underlines this aspect very clearly:

> there was no valley, nor hollow, no hill, no mound, no shadow of hill nor of mound by which I could mark the way I was making. Hour by hour I advanced, and saw no change – I was still the very centre of a round horizon; hour by hour I advanced, and still there was the same, and the same, and the same . . .[14]

The sun of course blazes, the air lies dead, everything is still and lifeless, and Kinglake falls asleep. He is awakened 'by a peal of church bells – my native bells – the innocent bells of Marlen [the local name for St Mary Magdalen, Taunton], that never before sent forth their music beyond the Blaygon hills!' Kinglake rouses himself, drawing 'aside the silk that covered my eyes and plunging my bare face into the light'. But still the bells ring on for, he estimates, another ten minutes. A mixture of rationality and nostalgia then compete as he attempts an explanation which

> attributed the effect to the great heat of the sun, the perfect dryness of the clear air through which I moved, and the deep stillness all around me [causing a] consequent susceptibility of the hearing organs [which] had rendered them liable to tingle under the passing touch of some memory that must have swept across my brain in a moment of sleep.

Conversely, while conceding that the people of Marlen are 'unaddicted to the practice of magical arts', he imagines that at the very moment he experienced this auditory mirage, 'the church-going bells of Marlen must have been actually calling the prim congregation of the parish to morning prayer'.[15] Something like Kinglake's Somerset domestic nostalgia also appears in the ostensibly exotic extravagance of Abbott's Malwa Fata Morgana near the beginning of my account (illus. 1 and 2). In the fanciful lithographs claiming to depict visions in central India, we can see church spires and gothic ruins that could easily be confused with Gilpin's depiction of Tintern Abbey in *The River Wye*.[16] Abbot's attractive account might be seen as a conversion of those optical qualities conventionally associated, since the late eighteenth century, with the Burkean Sublime into the Picturesque. In other words, visual experiences that might otherwise be threatening and opaque are incarnated as agreeable and enchanting, and rendered assimilable through ornamental gothic ruins.

There are similar 'sound mirages' in Hergé's Tintin adventure *Land of Black Gold* (first published in 1939), for, as Tom McCarthy has noted, 'the desert's visual field dissolves into orphaned sounds and disembodied voices', which he likens to Barthes' vision (in *S/Z*) of literature as a 'stereographic' space in which utterances are de-originated.[17] But in Tintin there is a domestication reminiscent of Kinglake in which the mirage becomes the subject of comedy, as Thompson and Thomson dive into the dry desert in swimsuits worthy of the English seaside postcard artist Donald McGill.

If Kinglake established the conditions for an auditory mirage, we might say that Bob Hope and Bing Crosby established the conditions for a gustatory mirage in the film *Road to Morocco* (1942). Lost in the desert, they see a drive-in mirage ('Herbert's Sandwiches'), where they anticipate getting hamburgers and beers. They run towards it shouting 'It's moving, we've got to grab it,' and then acknowledge its instability ('You see it, now you don't see it') before conceding its illusory nature but strong sensory presence ('Sure was a good one: I could even smell the onions'). This is a ludic demystification of the Fata Morgana, a dose of American scepticism about European Orientalist travellers' tales, but it powerfully territorializes and domesticates mirage.

Conversely, a liberating de-territorializing trajectory might be seen in the mirage as a prototype for imaginary visions, what the Theosophists Annie Besant and Charles W. Leadbeater would call 'thought forms'. These were positive synaesthetic forms that embraced illusion and imagination not as sources of error but as portals into a new realm of Theosophist perception. Thought forms, existing on the border between body and spirit, were deemed ultimately capable of capture through photography. Besant and Leadbeater point to Röntgen's rays, which have 'rearranged some of the older ideas of matter', and to Baraduc's experiments in thought-photography, evidence of the permeable boundary of the visible.[18] In Besant and Leadbeater's synthesis of Hindu and Muslim mysticism, we can see the early signs of how the 'East',[19] enduring in the 'blind-spot' of European science, would be transformed in the Western imagination from a site of corrupting Nabobism or of Ottoman decay in which illusion was a negative visual poison into an energizing synaesthetic realm in which refraction and the imagination were repurposed as part of a new, ostensibly positive, neo-Romantic project. This could of course also be seen as an echo of a Romantic narcotic discourse, which from De Quincey onwards had always shadowed the entire history of mirage.

We should also recognize the ways in which an increasingly connected global world, and new media flows and platforms, position mirages within different times and spaces. The 'traveller's report' in which exotic mirages were offered (several years after the event) to the metropolitan cognoscenti is long gone. That relatively stable world, with its slow travel and elongated passage from manuscript into letterpress, appears infinitely remote. Mirages now inhabit new interconnected horizons and provoke immediate responses.

The disintegration of that earlier mirage-scape is perhaps presaged by a strange coincidence in northern Pakistan. On 2 May 2011 a helicopter-borne assault force left Afghanistan en route for a large house in a compound less than a mile from the Pakistani Military Academy. The town, Abbotabad, had been founded in 1853 following the annexation of Punjab by one Major James Abbott, whom we encountered at the beginning of this narrative beholding a magnificent Fata Morgana in Malwa. The compound's tenant, who would be shot and killed during the raid, was Osama bin Laden, a connoisseur of dream interpretation.

Bin Laden and many of his followers, together with sundry other *jihadis* (as has been well documented by Iain Edgar and Amira Mittermaier, among others), are as committed to the analysis of dreams as were Ibn Khaldun and the employers of Kadare's Mark-Alem in *The Palace of Dreams*. Edgar sets recent accounts of Taliban and Al-Qaeda dream experiences against the surer footing of a deep Islamic tradition of attending to the dreams as political prophecy. Muhammad's 'night journey' (the *Lailatal-mi'raj*, the *mi'raj* that is not a mirage), in which he travels from Mecca to Al-Aqsa in Jerusalem to understand the secrets of the cosmos, is both dream and actual journey, and without doubt a privileged source of knowledge. The dream as political prophecy within Islam builds upon the Hadith's account of Muhammad's own immersion in the oneiric,and early theorists such as the ninth-century Ibn Ishaq al-Kindi and Ibn Khaldun.[20]

Mullah Omar of the Taliban had frequent dreams that formed the basis for his military planning, and Osama bin Laden himself is reported to have expressed anxiety that his followers' anticipatory dreams of the World Trade Center attack made his plan vulnerable.[21] The revelations within these dreams occupied a privileged position within heirognosis, the 'hierarchical classification of the different orders of knowledge displayed both in dreams and waking realities'.[22] But from our point of view, the chief interest attaches to the way in which dream theory occupies part of a spectrum in which mirages also play a role. Edgar links al-Kindi's notion of 'form-creating faculty' (outlined in his *Epistle on the Nature of Sleep and Dream*) to Sufi concepts of the 'imaginal' (*Alam al-mithal*) world. The 'imaginal' is quite distinct from the 'imaginary', since it is not unreal. Rather, the imaginal describes a world 'between that of sensibility and intelligibility . . . defined as "a world of autonomous forms and images"'.[23]

The events of 9/11 and the execution of bin Laden are two instances that reveal the manner in which different ideologies and lifeworlds collide within a landscape of fast media. Within this landscape, 'civil society' has been remade through the millions that have access to the Internet. The Parisian coffee shop has been replaced by the blogosphere and the 'metronome of apocalyptic time'.[24] Some of

43 The Foshan Fata Morgana, 20 October 2015. The Chinese banner on CTI Chung T'ien News's coverage reads 'A Fata Morgana in Foshan? Netizens suspect it is a glass reflection', suggesting scepticism about the status of the image.

the consequences of this new space of interpretation are indicated by the response to a spectacular Fata Morgana that appeared in October 2015 above Foshan in Guandong, China (illus. 43). The image took the form of looming dark tower blocks and by all accounts appeared for only a few minutes, during which it was extensively filmed. Internet conspiracy chat immediately connected it to Project Blue Beam, the secret NASA programme publicized by the late Quebecois conspiracy theorist Serge Monast in the early 1990s. Monast claimed that Blue Beam aimed to create a new world order overseen by the Antichrist. The Foshan Fata Morgana was taken by some as evidence of stage two of Monast's seven-stage prediction, which he termed 'The Big Space Show in the Sky'. I quote from Monast's central paranoid text:

> The second step in the NASA Blue Beam Project involves a gigantic 'space show' with three-dimensional optical holograms and sounds, laser projection of multiple holographic images to different parts of the world, each receiving a

> different image according to predominating regional national religious faith.

Monast's increasingly fervid proclamation ends with a Tocquevillean warning:

> If you cannot see, *if you cannot learn*, if you cannot understand, then you and your family and friends will succumb to the fires of the crematoria that have been built in every state and every major city on earth, built to deal with you.[25]

Conspiracies are mobile and mirages continue to generate new alliances with politics, which still attest to the central connection with the question of spectatorial democracy. Monast offers us an apocalyptic echo of Tocqueville's observation that democratic citizens

> like to discern the object which engages their attention with extreme clearness . . . they rid themselves of whatever conceals it from sight, in order to view it more closely and in the broad light of the day. This disposition of mind soon leads them to condemn forms which they regard as useless or inconvenient veils between them and the truth.[26]

Monast and the new attention given to his ideas by the Foshan Fata Morgana suggest that the power of the mirage is undiminished. Waterless seas, and cities in the sky, continue to act as zones of exploration and critique.

# Epilogue: Real, But Not True

Like the traveller in the desert, or the early mariner, I have had to contend with unexpected events, twists, turns, crises and astonishments. This has been one of the benefits of travelling through a terrain in which so much remains to be discovered and mirages fold into visions and dreams. It is, in part, because science has not yet 'killed' the mirage: the mirage still has a certain autonomy, sheltering in the blind spot of the not yet fully rationalized. This is on the one hand because of a puzzling lack of interest in mirages, but perhaps it is also because they continue to operate as zones of blindness and insight, or what we might call translucent occlusion (what Hartwig called a 'wonderful deception of vision').[1] We might see evidence here of the enduring attraction of the Sublime potential of 'obscurity'. In the Burkean formulation, occlusion has a paradoxical positivity: 'It is our ignorance of things that causes all our admiration, and chiefly excites our passion.'[2]

Science's puzzling disinterest in mirages is indeed puzzling. In 1961 the America correspondent of the British magazine *New Scientist* underscored how remarkably little was known about them.[3] This report triggered a series of letters and a commissioned piece by the literary scholar Richard Beck suggesting that Milton may well have attempted to describe mirages, and probably largely understood their physics, but that he lacked the word to describe it.[4] This prompts Beck to express his astonishment at 'what must surely be one of the most inexplicable blind-spots in the history of science. Years after the last witch had been burned and the *ignis fatuus* had been rationalized into methane, the mirage remained without an explanation and even without a name!'[5] It is astonishing, but perhaps not inexplicable, for what many of the accounts we have examined here show clearly is that, as Edmund Burke diagnosed so well, visual occlusion, opacity and mystery are entangled with the Sublime. They contain too many perverse pleasures to be surrendered prematurely to science. I. A. Richards once complained that doing practical criticism (of a poem, for instance) was like 'slitting the throat of a nightingale to see what makes it sing'. Because science has never fully wielded its knife, the mirage continues to sing in an obscure and mysterious manner.

A theme that has run throughout this account concerns mirage as a metaphor and as part of a moral lesson. Mirage, metaphorically speaking, is what is 'not there', what it is illusory, what is 'false'. Much the same position can, as we have seen, be found in the Koran. Mirage plays a similar role in Christian discourse; it signifies what is not 'really' there: it is a trap, an illusion. William Haig Miller's popular text from the 1860s has as its frontespiece an Arab camel driver, depicted by John Tenniel, staring at a not-too-distant Fata Morgana. However, the bulk of the book is directed at a domestic British and Christian audience (it was published by the Christian Tract Society) with a message concerning the pursuance of 'false and illusive streams'. The book is a Christian warning against the 'allurements of the world', especially the pursuit of pleasure, ambition and wealth.[6]

However, mirages are not simply 'false and illusive'. An early contribution to *Scientific American* pointed out that mirages cannot 'be attributed wholly to the exercise of the imagination', and indeed many modern mirage commentators would argue that they cannot be accurately termed optical 'illusions'.[7] As James H. Gordon put the matter, the 'belief that a mirage is something unreal, a sort of trick played on the eye, is wrong. The picture a mirage presents is real but never quite accurate.'[8] Mirages produce real effects on sense perception. They are real optical phenomena and of course they can be photographed. They may not be 'true', but they are certainly 'real'. The Cambridge clergyman and antiquarian Edward Clarke described a journey at the end of the eighteenth century to Rosetta in Egypt. He sees a mirage between his caravan and the city that is their destination: water appears to block the way forward. He stresses his certainty ('Not having in my own mind at the time, any doubt as to the certainty of its being water'[9]), which he shares with his Greek interpreter. His Arab guides assure him that they will reach Rosetta within an hour if they continue in a straight line, forcing him to express his exasperation: 'do you suppose me an idiot to be persuaded contrary to the evidence of my senses?' At this point, Clarke mordantly concludes, the guides bid him and his interpreter look behind them and see that they have walked through a great watery mirage:

> The view of it afforded us ideas of the horrible despondency to which travellers must sometimes be exposed, who, in traversing the interminable Desert, destitute of water, and perishing with thirst, have sometimes this deceitful prospect beyond their eyes.[10]

Mirages give a compelling account of the world because they are optically real. They are not hallucinations or illusions, but are rather features of the physical environment. They are also, of course, obscure and confusing. In this respect, they might be thought of as an example of 'Baroque' vision. Christine Buci-Glucksmann, whose interpretation has been championed by Martin Jay, crucially stresses the Baroque image's 'anti-Platonism'. The term 'Baroque' is thought to be etymologically derived from the Portuguese term for a misshapen pearl, its voluptuous tain creating complex refractive effects. It is, Jay suggests, 'Anti-Platonic in its disparagement of clarity and essential form, baroque vision celebrated instead the confusing interplay of form and chaos, surface and depth, transparency and absurdity'.[11] To this we could add Gilles Deleuze's celebration of the Baroque's 'stretched canvas diversified by moving, living folds' and Leibniz as the prophet 'of curves and twisting surfaces'.[12]

This opposition between on the one hand clarity and transparency, and on the other confusion and occlusion, recalls a similar distinction made by the 'Neuroarthistorian' John Onians. Deserts, especially those of West Asia, were the perfect habitat for mirages. They were what Onians has called 'confused' environments. Whereas in Greece an object-rich environment which could be apprehended through clear air 'gave people a feeling that seeing was a relatively unproblematic experience', the space between Basra and Cairo (the route the great optical scientist Al-Haytham traversed) presented an environment of incompletion and deception: 'objects were frequently absent and when they were present were likely to be concealed by heat haze or sandstorm [and] vision was made uncertain.'[13] Onians's stark contrast points towards a location where visions, dreams and mirages have always been deeply entangled.

In fact, although Al-Haytham wrote very extensively about refraction, he has nothing to say (at least in his *Kitab al-Manazir* or *De Aspectibus*) about mirage.[14] This surprising fact is one of the ways in which mirage refuses to be pigeonholed, and should encourage us to fight any such 'territorialization'. Onians's stark distinction between the 'clear air' of Greece and the West Asian environment 'notorious for its visual and mental deception' combines elements of Tocqueville's concerns (Greece is of course the home of democracy) with later Orientalist preoccupations.[15]

The mirage's intransigence, its refusal to explain itself, and disenchantment's reluctance to force it to do so, may also be testament to an unresolved question in the history of Western metaphysics concerning the relationship between what is real and what is true. For Plato, what was true was eternal and incorporeal, the Idea, that presence inscribed on the soul, whose intentional meaning could be cross-checked through Socratic dialogue rather than being dispersed to the four winds by a writing.[16]

In *The Republic*, Plato had castigated the 'imitator', the 'creator of the phantom, who knows nothing of reality', suggesting that imitation was like carrying a mirror everywhere which allowed images to be created that represented objects 'as they appear to be. But not . . . as they truly are.'[17] The image conveys resemblance, but not form or character: 'Something *like* the real thing, but not itself the real thing.' Both the painter's image of a couch, and a couch made by a carpenter (to take Plato's example), would both be 'rather shadowy by comparison with truth'.[18] *The Republic* also offered the extended allegory of prisoners in a cave where 'what people . . . take for truth would be nothing more than the shadows of manufactured objects'.[19] In the *Theaetetus* we hear that 'This "appearing" or "seeming" without really "being" . . . all these expressions have always been and still are deeply involved in perplexity.'[20] But mirages reconfigure (or even better 'refract') Plato's hierarchy between essence (superior) and appearance (inferior), or original and copy. The mirage, indeed, would be a perfect example of what Deleuze calls a 'simulacrum', that thing which 'is not a degraded copy. It harbors a positive power that denies *the original and the copy. The model and the reproduction*.'[21] What does a mirage 'copy'? Nothing: it is in this sense profoundly egalitarian, attacking the hierarchy between 'original' and 'copy'. It is a function of temperature gradients or inversions that disorder human expectations of representation. It offers a lesson diametrically opposed to the magical foundation myths of mimesis. It abolishes 'likeness' and 'resemblance'. It conjures, as Martin Swayne put it, 'a country of topsy-turveydom as regards the subjective estimate of the eyes'.[22] It is the uncanny, the *unheimlich*, the reliable mark of alterity, of not being at home (Kinglake notwithstanding), of no longer being able to invest in 'autopticism', 'ocularity' and all those other crutches of European epistemology, those proofs of visual familiarity and evidential certainty.

The mirage offers a different way into Plato's problem (and we might say into most of the subsequent history of Western philosophy), for it proffers a sense experience independent of the idea, an image that is freakishly beyond 'representation'. 'Simulation', as Deleuze puts it, 'is an image without resemblance'; it 'is the phantasm itself'.[23] It always takes its beholders by surprise, hence all those 'halted travellers' caught in the act of wonder. The mirage can certainly be a mirror: we read into it what we want, politically and culturally, but it appears to be a universally grasped optical effect. Real, but not true. That is the profound and inextinguishable lesson of the mirage.

# GLOSSARY

*Al-Haytham*: Arab experimental physicist, *c.* 965–1040 CE. Born in Basra but spent most of his life in the Fatimid Caliphate in Cairo. A pioneer in the study of reflection and refraction, often considered the father of optics. Also known as Alhazen, or Alhacen.

*Fata Morgana*: a complex superior mirage often seen in the Strait of Messina, believed by some to be 'fairy castles' created by the Arthurian sorceress Morgan le Fay to lure sailors to their death.

*ghost riders*: apparition of horse riders in the sky in the 'Great American Desert', often inverted, as with the Seventh Cavalry prior to the Battle of Little Big Horn in 1876. Made famous in literature by Charles Fenno Hoffman in the 1830s.

*inferior mirage*: an illusion seen below the real object, such as a lake seen in the desert (a refraction of the blue sky above).

*Kaaba*: black granite cube in the Grand Mosque, Mecca, on seeing which in 1853 Richard F. Burton writes about the 'mirage medium of Fancy'.

*mi'raj*: the Prophet Muhammad's night journey.

*mrigtrishna*: Sanskrit term referring to 'the deer's longing' or 'the thirst of gazelles'. Deer search for water in the desert before meeting their death under a blazing sun.

*Penglai/Horai*: a 'phantom paradise', or mirage island, in the China Sea.

*sehrab/suhrab/sarāb*: 'water of the desert', the Arabic term for mirage.

*shinkiro*: a mirage, which Japanese tradition ascribes to the exhalation of clams.

*superior mirage*: effect of temperature inversion, common in Arctic areas. Warm air above colder air causes the image to appear above the true object.

*Tabir Sarrail*: the Ottoman institute dedicated to the study of premonitions in Ismail Kadare's novel *The Palace of Dreams*.

*waterless sea*: Richard F. Burton's translation from *bahr bila ma*, an Arabic term for the Sahara, 'the sea without water'.

# REFERENCES


## Prologue: Chasing Mirage

1 Henry Miller, *The Colossus of Maroussi* (London, 1942), p. 29.
2 A. R. Wallace, 'Misleading Cyclopaedias', *Nature* (28 November 1872), p. 68.
3 Borges's imperial cartographers appeared in the one-paragraph-long short story 'On the Rigour of Science', first published in *Los Anales de Buenos Aires*, I/3 (March 1946).

## 1 Strange Visions Under a Cliff in Central India, October 1829

1 James Abbott, 'On the Mirage of India', *Journal of the Asiatic Society of Bengal*, XXIII (1854), p. 165.
2 Ibid., pp. 163, 164.
3 Ibid., pp. 165–6.
4 Ibid., p. 166.
5 Ibid., p. 168.
6 Ibid., p. 166.
7 Chumbul and Jumna are Chambal and Yamuna. James Tod, *The Annals and Antiquities of Rajasthan or the Central and Western Rajpoot States of India* (Calcutta, 1899), vol. I, p. 786.
8 Ibid., p. 787.
9 M. Minnaert, *The Nature of Light and Colour in the Open Air* (New York, 1954), p. 49
10 C. Fitzhugh Talman, 'The Real Fata Morgana: What is Known To-day About the Famous Phantoms of the Calabrian Coast', *Scientific American* (13 April 1912), p. 347.
11 Tod, *Annals*, p. 787.
12 Ibid., p. 786.
13 P. V. Kane, *History of Dharmasastra (Ancient and Medieval Religious and Civil Law in India)* (Poona, 1962), vol. V, part 2, p. 763.

14 W. Crooke, *The Popular Religion and Folklore of Northern India* [1896], revd edn (Delhi, 1978), vol. I, p. 56.
15 Ibid., p. 57.
16 Marina Warner, *Phantasmagoria: Spirit Visions, Metaphors, and Media into the Twenty-first Century* (Oxford, 2006), p. 95. Crooke documents many other Indian stories of underwater palaces (*Popular Religion*, pp. 56–7). Talman suggests the legend was carried by Normans to southern Italy in the eleventh century. Morgan le Fay lured mariners to their destruction: 'the seamen would mistake the aerial city for a safe harbor, and would be led hopelessly astray in endeavoring to reach it.' Talman, 'The Real Fata Morgana', p. 345.
17 Ibid.; the later accounts include George Harvey's early nineteenth-century *Encyclopaedia Metropolitana* (London, 1845), vol. XXII, p. 191.
18 Ibid., p. 335.
19 Ibid.
20 Anonymous, *Die Wunder der Natur* (Berlin, 1913).
21 See Joscelyn Godwin, *Athanasius Kircher's Theatre of the World* (London, 2009), p. 208.
22 Cited by Henry Swinburne, *Travels in the Two Sicilies in the Years 1777, 1778, 1779 and 1780* (London, 1803), p. 366.
23 Talman, 'The Real Fata Morgana', p. 345.
24 Ibid., p. 346.
25 Ibid., p. 335.

## 2 A World History of Mirages: The Thirst of the Gazelle

1 'Sand-bordered hell, where the mirage flickers day long above the Bitter Lake' is Rudyard Kipling's description of Port Said in *The Light That Failed* (Leipzig, 1891), p. 30.
2 A tradition opposed by other traditions within Islam. See Amira Mittermaier, *Dreams that Matter: Egyptian Landscapes of the Imagination* (Berkeley, CA, 2011).
3 Alexander von Humboldt, *Views of Nature; or, Contemplations on the Sublime Phenomena of Creation* (London, 1850), p. 15.
4 Ibid., pp. 137–8. Much the same material is also cited by David Brewster, *Letters on Natural Magic Addressed to Sir Walter Scott, Bart* (London, 1834), p. 128.
5 Our contemporary use of the word 'cosmic' is indebted to Humboldt.
6 *Mrig* = deer; *trishna* = longing, craving or lusty desires.
7 See Akshaya Mukul's comprehensive and fascinating account of the cultural and religious motivations of the publishing house that produced *Kalyan*: Akshaya Mukul, *Gita Press and the Making of Hindu India* (Delhi, 2015).

8 *The Ramayan of Valmiki*, trans. Ralph T. H. Griffiths (Varanasi, 1963), pp. 277, 278.
9 F. W. Bain, *Bubbles of the Foam* (London, 1912), p. xi.
10 Ibid.
11 Ibid., p. xii.
12 Ibid.
13 With 'Mirage! mirage!', Bain is presumably here echoing David Brewster's description in *Letters on Natural Magic* of the auspicious shout 'Morgana, Morgana!', uttered in the Straits of Messina whenever 'exulted' observers spot this 'lucky phenomenon' (Brewster, *Letters on Natural Magic*, p. 134); Bain, *Bubbles*, p. xvi.
14 Wendy Doniger O'Flaherty, *Dreams, Illusion and Other Realities* (Chicago, IL, 1986), p. 268.
15 Ibid., p. 273.
16 Ibid. The *Yogavasistha* continues: 'he made a house of air in the sky in order to rule over the air and the sky. But after a while the house faded away. He cried out: "Oh, my house made of space, where have you gone?" and he built another, and another, and another, and all of them dispersed into the air.'
17 Cited by Doniger O'Flaherty, *Dreams*, p. 274.
18 Ibid., p. 268.
19 *Hamlet*, III.2. cited by Doniger O'Flaherty, *Dreams*, p. 366.


## 3 'Fallacious Evidence of the Senses'


1 Walter Benjamin, 'Little History of Photography', in *Walter Benjamin: Selected Writings*, vol. II, part 2: *1931–1934*, trans. Rodney Livingstone et al., ed. Michael W. Jennings, Howard Eiland and Gary Smith (Cambridge, MA, 1999), p. 510.
2 Daniel Rock, attrib., 'Fallacious Evidence of the Senses', *Dublin Review*, III (1837), p. 548.
3 Ibid., p. 541.
4 Richard Rorty, *Philosophy and the Mirror of Nature* (Princeton, NJ, 1980); Iwan Rhys Morus, 'Illuminating Illusions, or, the Victorian Art of Seeing Things', *Early Popular Visual Culture*, X/1 (February 2012), p. 38.
5 Brewster, cited by Morus, 'Illuminating Illusions', p. 40
6 Ibid., p. 37.
7 Ibid., p. 39.
8 Rock, 'Fallacious Evidence', p. 548.
9 'Viera', cited ibid., p. 526.
10 Ibid., p. 527.
11 Ibid.

12 Ibid., p. 528. Washington Irving provides a very similar account, based on the same sources, as 'Appendix XXV – The Imaginary Island of St Brandan', in *The Life and Voyages of Christopher Columbus* (London, 1876), vol. II, pp. 876–81. Irving concludes by agreeing with Father Feyjoo that the cause of all these sightings was 'certain atmospherical deceptions, like that of the Fata Morgana' (p. 881).
13 Rock, 'Fallacious Evidence', pp. 548–9.
14 Samuel Hibbert, *Sketches of the Philosophy of Apparitions; or, an Attempt to Trace Such Illusions to Their Physical Causes,* 2nd edn (Edinburgh, 1825), p. 16.
15 C. J. Andersson, *Lake Ngami: Explorations and Discoveries During Four Years' Wanderings in the Wilds of South Western Africa* (London, 1856).
16 'A Reported Great Lake in Africa Nowhere', *Scientific American* (13 December 1856), n.p.
17 W. H. Lehn and I. Schroeder, 'The Norse Merman as an Optical Phenomenon', *Nature*, 289 (29 January 1981), pp. 362–6.

## 4 'Mocking Our Distress'

1 'Optical Marvels in the Antarctic: Light-pillars, Coronas, Auroras and Mirages that Greet the Southern Explorer', *Scientific American Supplement*, 1991 (28 February 1914), p. 132.
2 *Byron's Journal of his Circumnavigation, 1764–1766*, ed. Robert E. Gallagher (Cambridge, 1964), p. 30.
3 R. R. Madden, *Travels in Turkey, Egypt, Nubia, and Palestine, in 1824, 1825, 1826, and 1827* (London, 1829), vol. II, p. 200.
4 Henry Pottinger, *Travels in Beloochistan and Scinde; Accompanied by a Geographical and Historical Account of Those Countries* (London, 1816), p. 185.
5 'The Mirage', *Penny Magazine* (25 January 1834), p. 29.
6 E. A. Reynolds-Ball, *Practical Hints for Travellers in the Near East* (London, 1903), p. 41.
7 J. E. Richter, 'Effects of Thirst', *Scientific American Supplement*, 227 (8 May 1880), p. 3619.
8 F. L. James, *The Unknown Horn of Africa: An Exploration from Berbera to the Leopard River* (London, 1888), p. 95.
9 Murray Leeder, 'M. Robert-Houdin Goes to Algeria: Spectatorship and Panic in Illusion and Early Cinema', *Early Popular Visual Culture*, VIII/2 (May 2010), p. 218.
10 See Rachel O. Moore, *Savage Theory: Cinema as Modern Magic* (Durham, NC, 1999); Michael Tausig, *Mimesis and Alterity: A Particular History of the Senses* (New York, 1993); and Christopher Pinney, *Photography and Anthropology* (London, 2011).

11 Simon During, *Modern Enchantment: The Cultural Power of Secular Magic* (Cambridge, MA, 2004), p. 2, emphasis added; Leeder, 'M. Robert-Houdin', p. 210.

## 5 Cold and Hot: The Geography of Mirage

1 William Scoresby, *An Account of the Arctic Regions with a History and Description of the Northern Whale Fishing* (Edinburgh, 1820), vol. I, p. 386.
2 Ibid., p. 384. For a later commentary, see P. G. Tait, 'State of the Atmosphere Which Produces the Forms of Mirage Observed by Vince and by Scoresby', *Nature*, 27 (24 May 1883), pp. 84–8.
3 William Edward Parry, *Journal of a Voyage for the Discovery of a North-west Passage from the Atlantic to the Pacific* (London, 1821), p. 174.
4 Sir John Ross, *A Voyage of Discovery Made Under the Orders of the Admiralty in His Majesty's Ships Isabella and Alexander for the Purpose of Exploring Baffin's Bay and Inquiring into the Probability of a North-west Passage* (London, 1819), p. 246.
5 Parry, *Journal of a Voyage*, p. 32.
6 Robert Huish, *Last Voyage of Captain Sir John Ross R. N. Knt. to the Arctic Regions* (London, 1836), p. 104.
7 'Optical Illusions', *Scientific American* (2 September 1868), p. 153.
8 Emily C. Wilson, 'Prairie Scene: Mirage', www.alfredjacobmiller.com/artworks/prairie-scene-mirage, accessed 15 March 2017.
9 Charles Fenno Hoffman, *Wild Scenes in the Forest and Prairie With Sketches of American Life* (New York, 1843), pp. 15–40; The narrative would be reworked later in Charles M. Skinner's *Myths and Legends of Our Own Land* (Philadelphia, PA, 1896). 'Nights in an Indian Lodge: From the Portfolio of a Western Tourist', *American Monthly Magazine*, V (1835), pp. 62–77.
10 'The Ghost-rider: A Legend of the American Desert', *Bentley's Miscellany*, IV (July 1838), p. 472. The *Bentley's* version was subsequently reprinted in *London and Paris Observer*, XIV (1838), pp. 776–9.
11 Ibid., p. 473.
12 Ibid., p. 475.
13 Ibid., p. 476
14 Ibid., p. 477.
15 Ibid., p. 482, capitalization in original.
16 Ambrose Bierce, 'The Mirage', in *Collected Works of Ambrose Bierce*, (New York, 1909), vol. I, pp. 374–5.
17 Libby Custer, George's wife, offers this sense of premonition in her memoir *Boots and Saddles*: 'From the hour of breaking camp, before the sun was up a mist had enveloped everything. Soon the bright sun began to penetrate this veil and dispel the haze, and a scene of wonder and beauty appeared.

The cavalry and infantry in the order named, the scouts, pack-mules, and artillery, and behind all the long line of white-covered wagons, made a column altogether some two miles in length. As the sun broke through the mist a mirage appeared, which took up about half of the line of cavalry, and thenceforth for a little distance it marched, equally plain to the sight on the earth and in the sky. The fate of the heroic band, whose days were even numbered, seemed to be revealed, and already there seemed a premonition in the supernatural translation as their forms were reflected from the opaque mist of the early dawn.' (Libby Custer, *Boots and Saddles: On Life in Dakota with General Custer* (New York, 1885)).

18 'Little Big Horn Associates Message Board', http://thelbha.proboards.com/thread/1825/libbys-mirage-7th-cavalry, accessed 12 March 2016.

19 John Lear, 'The Reality Behind Mirages', *New Scientist*, CCXIX (26 January 1961), p. 208.

20 'Atmospheric Illusions in India', *Chambers' Edinburgh Journal* (4 June 1836), p. 151.

21 Cited in 'The Wanderer in Syria', in *The New Quarterly Review and Digest of Current Literature, British, American, French, and German*, 1/3(July 1852), p. 253.

22 W. H. Bartlett, *Forty Days in the Desert on the Track of the Israelites; or, a Journey from Cairo to Mount Sinai and Petra* (London, n.d.), p. 9. See also 'Forty Days in the Desert', *Chambers' Edinburgh Journal* (23 December 1848), pp. 406–9.

23 'Mirage', *Chambers' Edinburgh Journal* (12 January 1895), p. 32.

24 'Evenings in My Tent', *Edinburgh Review* , C/204 (1854), p. 403.

25 James Richardson, *Travels in the Great Desert of Sahara in the Years 1845 and 1846 Containing a Narrative of Personal Adventures During a Tour of Nine Months through the Desert, among Touaricks and other Tribes of Saharan People* (London, 1848), vol. II, chap. 23.

26 'The Enchanted Bay', *Chambers' Edinburgh Journal* (17 February 1849), p. 111.

27 S. Dalí, *La Conquête de l'irrationnel* (Paris, 1935), illus. 30.

## 6 Mirage and Crisis

1 *The Works of Flavius Josephus*, trans. William Whiston (London, n.d.), p. 653.

2 *The Book of Common Prayer* (London, 1735), n.p.

3 Anon., *A Strange and True Relation of Severall Wonderful and Miraculous Sights* (London, 1661), transcribed in *Anomalous Phenomena of the Interregnum* (London, 1991), p. 26.

4 Ibid., p. 27.

## 7 Oriental Mirages and 'Spectatorial Democracy'

1 Beck notes that the *Dizionario etimologico italiano* refers to a much earlier French usage of *mirage*. The date proposed is 1753, a date which if true, he suggested, would 'tarnish the reputation for linguistic ingenuity of Napoleon's Egyptian expedition' (Richard J. Beck, 'Milton and Mirages', *New Scientist*, CCXXVI (16 March 1961), note at p. 692).
2 'On the Optical Phaenomenon of Mirage, translated from the French of M. Gaspard Monge, by the Author of "A Non-military Journal made in Egypt"', *New Annual Register; or, General Repository of History, Politics, and Literature* (January 1803), p. 187.
3 Beck would later note that the first usage of mirage in English was in 1803 in a paper in the *Philosophical Transactions of the Royal Society* by W. H. Wollaston, who would later write that 'I was not aware that an account had been given by M. MONGE of the phenomenon known to the French by the name of *mirage*, which their army had daily opportunities of seeing in their march through the deserts of Egypt.' Beck, 'Milton and Mirages', p. 692.
4 Georg Hartwig, *The Aerial World: The Phenomenon and Life of the Atmosphere* (London, 1875), p. 209.
5 'On the Optical Phaenomenon of Mirage', p. 192.
6 William Haig Miller, *Mirage of Life* (London, 1867), p. 14.
7 'Biography – Monge', *Edinburgh Annual Register* (January 1819), p. 333.
8 Charles M. Doughty, *Travels in Arabia Deserta* (London, 1921), p. 35.
9 T. E. Lawrence, *Seven Pillars of Wisdom: A Triumph* (London, 1935), p. 65.
10 Ibid., p. 304.
11 Ibid., pp. 575–6.
12 Ibid., p. 387.
13 Edward W. Said, *Orientalism* (London, 1978).
14 Chris Otter, *The Victorian Eye: A Political History of Light and Vision in Britain, 1800–1910* (Chicago, IL, 2008).
15 Yaron Ezrahi, 'Dewey's Critique of Democratic Visual Culture and its Political Implications', in *Sites of Vision: The Discursive Construction of Sight in the History of Philosophy*, ed. David Michael Levin (Cambridge, MA, 1999), p. 316.
16 Alexis de Tocqueville, *Democracy in America and Two Essays on America*, trans. Greald E. Bevan (London, 2003), p. 495.

## 8 From Clam-monsters to Representative Democracy

1 See www.biwa.ne.jp/~t-ban and http://koayu.eri.co.jp/biwadas/sinkirou/sinkirou.htm, the result of many years' observation and photography by Tadashi Ban and Kuzuyuki Matsui, Japanese physics teachers.

2 Edward H. Schafer, 'Fusang and Beyond: The Haunted Seas to Japan', *Journal of the American Oriental Society*, CIX/3 (1989), p. 392.
3 Ibid., p. 395.
4 Lafcadio Hearn, *The Romance of the Milky Way* (Boston, MA, and New York, 1907), p. 67.
5 Collection of the Museum of Fine Arts, Boston, 11.25783.
6 T. Shaw, *Travels, Or Observations Relating to Several Parts of Barbary and the Levant* (Oxford, 1738), p. 379.
7 Schafer, 'Fusang and Beyond', p. 395.
8 The aquarium stamp is dated year two of the Taisho era (1913); a handwritten inscription on the card is dated December 1917. Thanks to Keiko Homewood for her translation.
9 Lafcadio Hearn, *Kwaidan: Stories and Studies of Strange Things* (Boston, MA, and New York, 1904), p. 173.
10 Ibid., p. 178.
11 Timon Screech notes that *iro,* the Japanese term for 'colour', was also 'a Buddhist technical term for illusion'. Timon Screech, *The Lens Within the Heart: The Western Scientific Gaze and Popular Imagery in Later Edo Japan* (Honolulu, HI, 2002), p. 192.
12 See Barbara Celarent's enlightening review of Fukuzawa Yukichi's *An Outline of a Theory of Civilization*, trans. David A. Dilworth and G. Cameroon Hurst III, in *American Journal of Sociology*, CXIX/4 (2009), pp. 1213–20.
13 See Screech, *The Lens Within the Heart*, on the changing fortunes of 'downward vision'.

## 9 The Halted Viewer and *Sfumato*

1 *Harper's Monthly Magazine* (October 1860), p. 614.
2 Joseph Leo Koerner, *Caspar David Friedrich and the Subject of Landscape* (London, 1995), p. 233.
3 Ibid., p. 160.
4 Ibid., p. 164.
5 Michael Gaudio, *Engraving the Savage: The New World and Techniques of Civilization* (Minneapolis, MN, 2008), p. 53.
6 Ibid., p. 48.
7 Ibid., p. 59.
8 Mary W. Helms, *Ulysses' Sail: An Ethnographic Odyssey of Power, Knowledge, and Geographical Distance* (Princeton, NJ, 1988).
9 James H. Gordon, 'Mirages', in *Annual Report of the Board of Regents of the Smithsonian Institution* (Washington, DC, 1959), p. 343. Gordon, mysteriously, seems to mean this literally.

10 Wiliam M. Ivins, Jr, *Prints and Visual Communication* (Cambridge, MA, 1953).
11 Anthony Pagden, *European Encounters in the New World* (New Haven, CT, 1993), pp. 51–88.
12 'A City in the Air', *Scientific American*, (19 December 1846), p. 101.
13 'Splendid Mirage in Paris', *Scientific American* (14 August 1847), p. 370.
14 *English Mechanic*, LXVIII (1898), p. 375, available at http://aty.sdsu.edu. bibliog/bibliog.html. See also 'Photographic Progress in 1879' in *Chambers' Journal of Popular Literature* (15 May 1880), p. 311, which reports on the exhibition of the image at the Photographic Society of Great Britain: 'Here we have the inverted image of a gunboat in the sky, appearing above a church steeple; this was the effect of a mirage seen at Tenby.'
15 Franz Boas, *Franz Boas Among the Inuit of Baffin Island, 1883–84: Journals and Letters*, ed. Ludger Muller-Wiles, trans. William Barr (Toronto, 1998), p. 60. Boas had researched a PhD in physics in Kiel on the question of reflection, polarization and the colour of seawater. See S. Helmreich, 'Nature/Culture/Seawater', *American Anthropologist*, CXIII (2011), pp. 132–44, and George W. Stocking, 'From Physics to Ethnology', *Journal of the Hisorical Behavioral Sciences*, I (1965), pp. 53–66. I'm grateful to Simon Schaffer for leads relating to Boas.
16 Alexander Badlam, *Wonders of Alaska* (San Francisco, CA, 1890), p. 127.
17 Ibid., p. 131.
18 Ibid., p. 132.
19 Ibid., p. 133.
20 Ibid., pp. 135, 136.
21 Ibid., p. 137.
22 R. W. Wood, 'Mirage on City Pavements', *Nature*, 58 (20 October 1898), p. 596.
23 'The "Telephot", A Novel Apparatus for Photographing at Great Distance', *Scientific American* (27 June 1903), p. 486.
24 Anon., 'Optical Marvels in the Antarctic: Light-pillars, Coronas, Auroras and Mirages that Greet the Southern Explorer', *Scientific American Supplement*, 1991 (28 February 1914), p. 132.
25 Anon., 'Stereoscopic Photographs', *Scientific American Supplement*, 79 (23 January 1915), p. 51.
26 'The Photographic Mirage', *Scientific American Supplement*, 883 (3 December 1892), p. 14110.

## 10 Memory and Modernity

1 See 'Chinese Mirages', *Chambers' Edinburgh Journal* (10 July 1847), pp. 28–30, for an early discussion of mirage as a metaphorical, rather than solely optical, concept.

2 M. R. James, *Ghost Stories of an Antiquary* (Harmondsworth, 1937), p. 51.

3 Joseph Addison, 'Primary Pleasures: Effects on the Imagination from Nature', *The Spectator* (24 June 1712), repr. in Richard Steele and Joseph Addison, *Selections From The Tatler and The Spectator*, ed. Angus Ross (Harmondsworth, 1988), p. 376. Thanks to Jonathan Lamb for pointing out the importance of this text.

4 See Vincent LoBrutto, *Becoming Film Literate: The Art and Craft of Motion Pictures* (Westport, CT, 2005), p. 179.

5 The filmic realization builds upon T. E. Lawrence's own intensely 'cinematic' descriptions (for example, 'Through the mirage of heat which flickered over the shining flintstones of the ridge we could see, at first, only the knotted brown mass of the column, swaying in the haze. As it grew nearer the masses used to divide into little groups . . .'; *Seven Pillars of Wisdom: A Triumph* (London, 1935), p. 587.

6 An Atlantic mirage first described by the seventeenth-century buccaneer William Ambrosia Crowley.

7 As renamed by Richard Collinson, Captain of the HMS *Endeavour* on its 1850s voyage through the Bering Strait. See William Barr, *Artic Hell-ship: The Voyage of HMS Enterprise 1850–1855* (Edmonton, 2007), p. 93.

8 J. R. Jackson, *What to Observe; or The Traveller's Remembrancer* (London, 1841), pp. 104–5.

9 Sydney B. J. Skertchly, 'A Vivid Mirage', *Nature*, 2 (25 August 1870), p. 337.

10 Arthur E. Brown, 'Mirages', *Nature*, 1 (9 January 1890), p. 225.

11 L. N. Norris-Rogers and Harry Hillman, 'Mirage Effects', *Nature*, 104 (12 February 1920), p. 630.

## 11 Theatrical Mirages

1 Yaron Ezrahi, 'Dewey's Critique of Democratic Visual Culture and its Political Implications', in *Sites of Vision: The Discursive Construction of Sight in the History of Philosophy*, ed. David Michael Levin (Cambridge, MA, 1999), p. 316.

2 Finbarr Barry Flood, 'Correct Delineations and Promiscuous Outlines: Envisioning India at the Trial of Warren Hastings', *Art History*, XXIX/1 (2006), p. 49. Here Flood draws upon arguments advanced earlier by Sara Suleri.

3 Ibid., pp. 57, 51.
4 Ibid., p. 54.
5 *The Grand Master, or Adventures of Qui Hi? In Hindostan. A Hudibrastic Poem in Eight Cantos* (London, 1816). See Christina Smylitopolous, 'A Nabob's Progress: Rowlandson and Coombe's *The Grand Master*: A Tale of British Imperial Excess, 1770–1830', PhD thesis, McGill University (2010), and Zirwat Chowdhury, 'Imperceptible Transitions: The Anglo-Indianization of British Architecture, 1769–1822', unpublished PhD thesis, Northwestern University (2012).
6 The location of a major shrine depicting the Hindu *trimurti* near Mumbai, and which features iconographically in another of Rowlandson's images.
7 Salomo Friedlaender, 'Fata Morgana Machine', in Friedrich A. Kittler, *Gramophone, Film, Typewriter*, trans. Geoffrey Winthrop-Young and Michael Wutz (Stanford, CA, 1999), pp. 134–5.
8 Murray Leeder, 'M. Robert-Houdin Goes to Algeria: Spectatorship and Panic in Illusion and Early Cinema', *Early Popular Visual Culture*, VIII/2 (May 2010), p. 209.
9 Ibid., p. 217.
10 Graham M. Jones, 'Modern Magic and the War of Miracles in French Colonial Culture', *Comparative Studies in Society and History*, LII/1 (2010), p. 77.
11 Leeder, 'M. Robert-Houdin', p. 20.
12 Ibid., p. 210.
13 Simon During, *Modern Enchantments: The Cultural Power of Secular Magic* (Cambridge, MA, 2002), p. 143.
14 'Panorama of the Nile – Egyptian Hall', *The Mirror Monthly Magazine* (September 1849), p. 193.
15 Richard D. Altick, *The Shows of London* (Cambridge, MA, 1978), p. 245.
16 See Elizabeth Carlson, 'Reflections to Projections: The Mirror as a Proto-cinematic Technology', *Early Popular Visual Culture*, IX/1 (2011), pp. 15–35.
17 Zeynip Celik, *Displaying the Orient: Architecture of Islam at Nineteenth-century World's Fairs* (Berkeley, CA, 1992), p. 176.
18 Tom Gunning writes of the 'aesthetic of astonishment'. Boyer claimed that Hénard's intention was to 'bewilder the spectator as much as possible'; Jacques Boyer, 'The Palace of Mirages', *Scientific American* (15 May 1909), p. 371.
19 Ibid., pp. 371–2.
20 Carlson, 'Reflections to Projections', p. 18.
21 See Benjamin, *The Arcades Project*, trans. Howard Eiland and Kevin McLuaghlin (Cambridge, MA, 1999), pp. 537-42.
22 Carlson, 'Reflections to Projections', p. 20.

23 Ibid., p. 19.
24 Martin Jay, *Downcast Eyes: The Denigration of Vision in Twentieth-century French Thought* (Berkeley, CA, 1993); Jacques Lacan, *Four Fundamental Concepts of Psychoanalysis*, trans. Alan Sheridan (New York, 1978).
25 'The Hall of Mirrors: A World of Light', www.grevin-paris.com/en/univers/hall-mirrors, accessed 15 March 2017.
26 Boyer, 'The Palace of Mirages', p. 371.

## 12 The 'Mirage Medium of Fancy'

1 Richard Burton, *Personal Narrative of a Pilgrimage to Al-Madinah and Meccah* (New York, 1964), vol. II, p. 72.
2 Ibid.
3 'Singular Phenomena of Europe', *The Kaleidoscope; or, Literary and Scientific Mirror* (2 January 1821), p. 211.
4 G. Belzoni, *Narrative of the Operations and Recent Discoveries Within the Pyramids, Temples, Tombs, and Excavations in Egypt and Nubia* (London, 1820), p. 196. See also P. A. Macmahon, 'Mirage', *Nature*, 59 (12 January 1899), p. 260: 'One very curious thing about mirage is that it depends very much upon the position of the eye: a few inches in the height of the eye make all the difference. I remember myself, on the plains of India, observing a mirage which was only evident when I was at a particular height; there was only a vertical space of two or three inches in which the effect could be seen, so that these phenomena may easily escape notice.'
5 Augustus Ralli, *Christians at Mecca* (London, 1909), p. 161.
6 Ibid., p. 177.
7 Burton, *Personal Narrative*, p. 14.
8 Johan Ludwig Burckhardt (1784–1817) adopted Arab dress while studying at Cambridge before his journey to Aleppo and his rediscovery of the city of Petra. Ludovico di Varthema (*c.* 1470–1517) was the first non-Muslim to enter Mecca. Burton, *Personal Narrative*, vol. I, p. xx.
9 William Gifford Palgrave, *Narrative of a Year's Journey through Central and Eastern Arabia* (London, 1865), vol. I, pp. 258–9, cited by Burton, *Personal Narrative*, vol. I, p. xxi.
10 William Gifford Palgrave, *Narrative of a Year's Journey through Central and Eastern Arabia*, 5th edn (one vol.) (London, 1869), pp. 6, 199, emphasis added.
11 Burton, *Personal Narrative*, vol. II, p. 72.
12 Ibid., pp. 160–61.
13 Ibid., p. 99.
14 Joseph Packer, 'The Claim of Arabic Language and Literature' in *The Biblical Repository and Quarterly Observer* (1836), pp. 439–40.

Packer adds in a footnote that the mirage 'is a common emblem in Arabic poetry of disappointed hope'.

15 M. M. Badawi, *A Critical Introduction to Modern Poetry* (Cambridge, 1975), pp. 132, 135.

16 Jones notes that the English term 'mirage' names three quite distinct desert phenomena. He differentiates between true mirages (in Arabic *sarāb*), then 'shimmering' and finally what he calls 'gossamer'. Shimmering is produced by hot air rising into cooler air and gossamer is created by the filaments from plants whose seed pods burst as temperatures increase. Shimmering and gossamer evoke the liminal space between matter and atmosphere that we have already encountered in our consideration of *sfumato*. Jones then proceeds to examine these different optical effects in relation to a line from the *Lamiyyat al-'Arab* by the so-called sixth century 'brigand poet' (or *Sa'alik*), al-Shanfarā, which Jones memorably translates as 'Many a Dog-day, when the mirage melts away and the vipers slither over the sun-baked ground'. Alan Jones, *Early Arabic Poetry: Select Poems* (Reading, 2011), p. 197.

17 Elsaid Badawi and Muhammed Abdel Haleem, *Arabic-English Dictionary of Qur'anic Usage* (Leiden, 2008), pp. 874–75.

18 The title of Naguib Mahfouz/Najib Mahfuz's novel of 1948. See Philip Kennedy, 'Sons and Lovers and the Mirage', *Journal of Arabic Literature*, XLI (2010), pp. 46–65.

19 John Malcolm, *Sketches of Persia* (Philadelphia, PA, 1828), vol. II, p. 47.

20 Suzanne Conklin Akbari, *Idols in the East: European Representations of Islam and the Orient, 1100–1450* (Ithaca, NY, 2009), p. 250.

21 Ibid., p. 253.

22 Ibid., pp. 253–4.

23 For 'smoke' Burnes provides the Gujarati term *dhooan*. Alexander Burnes, *Travels into Bokhara: Being the Account of a Journey from India to Cabool, Tartary and Persia* (London, 1843), vol. III, pp. 320–21.

## 13 Mirage and Oriental Despotism

1 Jonathan Lamb, *Preserving the Self in the South Seas, 1680–1840* (Chicago, IL, 2001), p. 112.

2 Henry Miller, *The Colossus of Maroussi* (London, 1942), p. 42.

3 Ibid. Perhaps 'coryphantic' was an error on Miller's or his publisher's part and should be 'corybantic' (meaning frenzied).

4 Ibid., p. 43.

5 Much of this is well documented by Marina Warner in *Stranger Magic: Charmed Sttes and the Arabian Nights* (London, 2011).

6 Richard F. Burton, *The Book of the Thousand and One Nights: A Plain and Literal Translation of the Arabian Nights Entertainments. Reprinted from the*

*Original Edition and Edited by Leonard C. Smithers* (London, 1897), vol. I, p. xvii.

7 Ibid., p. xviii.

8 Ibid., vol. V, p. 9. Compare with Malcolm C. Lyons's recent Penguin Modern Classics translation: 'It is like the elusive dream of a sleeper, or a desert mirage which the thirsty man takes to be water and Satan embellishes it for man until the hour of his death.' (*The Arabian Nights: Tales of 1001 Nights* (London, 2010), vol. II. p. 525).

9 Romesh Chandra Dutt, *Arabian Nights*, trans. Richard F. Burton [1885] (New York, 2008), vol. IV. See William Barr, *Arctic Hell-ship: The Voyage of HMS Enterprise, 1850–1855* (Edmonton, 2007), p. 93.

10 Kate Douglas Wiggin and Nora A. Smith, eds, *The Arabian Nights: Their Best-known Tales*, illustrated by Maxfield Parrish (New York, 1909).

11 Ros Ballaster, *Fabulous Orients: Fictions of the East in England, 1662–1785* (Oxford, 2005).

12 Montesquieu, *The Persian Letters* (Harmondsworth, 1973), p. 66.

13 Ibid.

14 Ibid., p. 89.

15 Ibid., p. 158.

16 Ibid., p. 166.

17 Ibid., p. 181.

18 J.W.H. Stobart, *Islam and Its Founder* (London, 1876), p. 238.

19 'A Romance of the Mirage', *The Cornhill Magazine* (August 1883), p. 167.

20 Ibid., p. 168.

21 Ibid., p. 169.

## 14 Keeping Mecca and Medina Invisible

1 Ronald B. Inden, 'Orientalist Constructions of India', *Modern Asian Studies*, XX/30 (1986), pp. 429–32.

2 Arthur Rowan, *I Live Again: Travel, Secret Service and Soldiering in India and the Near East* (London, 1938), p. 197.

3 Augutus Ralli, *Christians at Mecca* (London, 1909), p. 1.

4 Ibid., p. 272.

5 Ibid., p. 273.

6 Ibid., p. 274.

7 Ibid.

## 15 Inside Abdul Hamid II's Head

1 Richard Davey, *The Sultan and His Subjects* (London, 1907), p. 83.
2 Ibid., p. 79.
3 Ibid., pp. 77, 78, 91.
4 Marina Warner, *Phantasmagoria: Spirit Visions, Metaphors, and Media into the Twenty-first Century* (Oxford, 2006), p. 95.
5 I am indebted to www-rohan.sdsu.edu/~aty/bibliog/bibliog.html for knowledge of this source.
6 Martin Swayne, *In Mesopotamia* (London, 1917), p. 66.
7 Ibid., pp. 66–7.
8 John Lear, 'The Reality Behind Mirages', *New Scientist*, CCXIX (26 January 1961), p. 208.
9 Ibn Khaldun, *The Muqaddimah: An Introduction to History*, trans. Franz Rosenthal, ed. N. J. Dawood (Princeton, NJ, 1969), p. 81.
10 Ibid., p. 82.
11 Ibid., p. 83.
12 Ismail Kadare, *The Palace of Dreams*, trans. Barbara Bray (London, 2008), pp. 9, 19, 21. Thanks to Kate Elizabeth Creasey for alerting me to this.
13 Ibid., pp. 38, 57.
14 Ibid., p. 99.
15 Robert A. Paul, '*CSSSH* Notes', review of Amira Mittermaier, *Dreams that Matter: Egyptian Landscapes of the Imagination*, *Comparative Studies in Society and History*, LVI/3 (2014), pp. 803–4.

## 16 Mirage *Pharmakon*: Wild and Domestic

1 Immanuel Kant, *Observations on the Feeling of the Beautiful and Sublime*, trans. John T. Goldwater (Berkeley, CA, 1960) p. 55. See the intriguing 'A Few Passages Concerning Omens, Dreams, Appearances &c.', *Blackwood's Edinburgh Magazine*, LVIII (December 1845), pp. 735–51.
2 Friedrich Kittler, *Optical Media: Berlin Lectures 1999*, trans. Anthony Enns, (Cambridge, 2010), p. 76.
3 C. Fitzhugh Talman, 'The Real Fata Morgana: What is Known To-day About the Famous Phantoms of the Calabrian Coast', *Scientific American* (13 April 1912), p. 346.
4 Kittler, *Optical Media*, pp. 77, 78, 80.
5 Ibid., p. 79.
6 The *pharmakon* is the term used by Plato in the *Phaedrus*, and catches Jacques Derrida's attention. Derrida notes how 'in the most striking manner the regular, ordered polysemy that has, through skewing, indetermination, or overdetermination, but without mistranslation, permitted the rendering

of the same word by "remedy", "recipe", "poison", "drug", "philtre", etc.' This double-ness, Derrida continues, 'is a difficulty inherent in its very principle, situated less in the passage from one language to another, from one philosophical language to another, than already . . . in the tradition between Greek and Greek.' Jacques Derrida, *Dissemination*, trans. Barbara Johnson (Chicago, IL, 1981), p. 71.

7 Georg Hartwig, *The Aerial World: The Phenomenon and Life of the Atmosphere* (London, 1875), p. 208.

8 There is perhaps a prefiguration of this in al-Biruni's 'assertion and disavowal of alterity'. See Finbarr B. Flood, *Objects of Translation: Material Culture and Medieval 'Hindu-Muslim' Encounter* (Princeton, NJ, 2009), p. 6.

9 Walter Benjamin, 'On the Mimetic Faculty', in *One Way Street and Other Writings*, trans. Edmund Jephcott and Kingsley Shorter (London, 1979), p. 160.

10 Hartwig, *The Aerial World*, p. 210.

11 Ibid., p. 211.

12 Roland Barthes, *The Pleasure of the Text* (New York, 1975), p. 60.

13 See Robert Springer, 'The Mysterious Sounds of Nature', *Popular Science Monthly* (October 1880), pp. 772–6, on 'sound mirages' including Kinglake's. Springer notes, 'The mirage of the *Fata Morgana* is sometimes accompanied with a sound like thunder' (p. 774).

14 Alexander Kinglake, *Eothen: Traces of Travel Brought Home from the East* (Oxford, 1982), pp. 184–5.

15 Ibid., p. 184.

16 William Gilpin, *Observations on the River Wye and Several Parts of South Wales &c. Relative Chiefly to Picturesque Beauty Made in the Summer of the Year 1770*, 3rd edn (London, 1792).

17 Tom McCarthy, *Tintin and the Secret of Literature* (London, 2006), n.p.

18 Annie Besant and Charles W. Leadbeater, *Thought Forms* (London, 1905), pp. 1–3.

19 See Natasha Eaton, *Colour, Art and Empire: Visual Culture and the Nomadism of Representation* (London, 2013), p. 252, on the Indian context of their work.

20 'Let us see, who had an interesting dream' he often asked. Iain R. Edgar, 'The Dream Will Tell: Militant Muslim Dreaming in the Context of Traditional and Contemporary Islamic Dream Theory and Practice', *Dreaming*, XIV/1 (2004), p. 22; for historical context, see Nile Green, 'The Religious and Cultural Roles of Dreams and Visions in Islam', *Journal of the Royal Asiatic Society*, XIII/3 (November 2003), pp. 287–313.

21 Edgar, 'The Dream Will Tell', p. 23.

22 Ibid., p. 25.

23 H. Corbin, cited ibid.

24 See J. M. Berger, 'The Metronome of Apocalyptic Time: Social Media as a Carrier Wave for Millenarian Contagion', *Perspectives on Terrorism*, IX/4 (August 2015).

25 Serge Monast, 'Project Blue Beam', http://educate-yourself.org/cn/projectbluebeam25jul05.shtml, 1994.

26 Cited Yaron Ezrahi, 'Dewey's Critique of Democratic Visual Culture and its Political Implications', in *Sites of Vision: The Discursive Construction of Sight in the History of Philosophy*, ed. David Michael Levin (Cambridge, MA, 1999), p. 316.

## Epilogue: Real, But Not True

1 Georg Hartwig, *The Aerial World: The Phenomenon and Life of the Atmosphere* (London, 1875), p. 208.

2 Edmund Burke, *A Philosophical Enquiry into the Origin of our Ideas of the Sublime and Beautiful* (Notre Dame, IN, 1968), p. 61.

3 It is salutary to find *New Scientist* at the start of the decade of the 'white heat of technology' describing mirages as a subject 'which almost everyone has heard about and about which almost nobody was well-informed' and that Gordon's 1959 report had established that 'mirages are real'. John Lear, 'The Reality behind Mirages', *New Scientist*, CCXIX (26 January 1961), p. 208.

4 Beck focuses on a passage from Book IV of *Paradise Regained* which explains how Satan may have managed to show Christ all the kingdoms of the world from a high mountain:

> By what strange Parallax or Optic skill
> Of vision multiplied through air, or glass
> Of Telescope, were curious to enquire (IV/40/2).

Beck's conclusion is that the blind Milton must have been aware of experiments with light (Richard J. Beck, 'Milton and Mirages', *New Scientist*, CCXXVI (16 March 1961), p. 693).

5 Ibid., p. 692.

6 William Haig Miller, *Mirage of Life* (London, 1867), pp. 19–20.

7 'Optical Illusions', *Scientific American* (2 September 1868), p. 153.

8 James H. Gordon, 'Mirages', in *Annual Report of the Board of Regents of the Smithsonian Institute* (Washington, DC, 1959), p. 328.

9 *Travels in Various Countries of Europe, Asia and Africa* (1810–19), vol. II, p. 295, cited by George Harvey, *Encyclopaedia Metropolitana* (London, 1845), p. 191, entry for 'Mirage'.

10 Cited ibid.

11 Martin Jay, *Downcast Eyes: The Denigration of Vision in Twentieth-century French Thought* (Berkeley, CA, 1993), p. 47.

12 Gilles Deleuze, *The Fold: Leibniz and the Baroque*, trans. Tom Conley (Minneapolis, MN, 1993), p. 27; Tom Conley, 'Translator's Foreword: A Plea for Leibniz', ibid., p. xi.
13 John Onians, *Neuroarthistory: From Aristotle and Pliny to Baxandall and Zeki* (New Haven, CT, 2009), p. 41.
14 See Mark A. Smith, trans., 'Alhacen on Refraction: A Critical Edition with English Translation and Commentary of Book 7 of Alhacen's De Aspictibus the Medieval Latin Version of Ibn al-Haytham's "Kitab al-Manazir"', *Transactions of the American Philosophical Society*, New Series, C/3 (2010). Smith observes that Al-Haytham was 'deeply concerned with "deceptive environments" and dealt at length with illusions caused by environmental factors such as haze, smoke . . . but didn't [at least in the *Kitab al-Manazir* or *De Aspectibus*] deal with optical phenomena arising from layers of varying air density' (personal communication, 19 September 2016). See also Hans Belting, *Florence and Baghdad: Renaissance Art and Arab Science*, trans. Deborah Lucas Schneider (Cambridge, MA, 2011), p. 92.
15 Onians, *Neuroarthistory*, p, 41.
16 R. Hackforth, 'The Superiority of the Spoken Word: Myth of the Invention of Writing', in *Plato's Phaedrus* (Cambridge, 1952), pp. 156–64.
17 Ravi Sundaram, 'The Pirate Kingdom Revisited', *Third Text*, XXIII (2009), p. 335.
18 Plato, *The Republic*, ed. G.R.F. Ferrari, trans. Tom Griffith (Cambridge, 2000), p. 315.
19 Ibid., p. 221.
20 F. M. Cornford, *Plato's Theory of Knowledge: The Theaetetus and the Sophist of Plato Translated with a Running Commentary* (London, 1960), p. 200.
21 Gilles Deleuze, *The Logic of Sense*, trans. Constantin V. Boundas, Mark Lester and Charles Stival (London, 2013), p. 271.
22 Martin Swayne, *In Mesopotamia* (London, 1917), p. 67.
23 Deleuze, *The Logic of Sense*, pp. 267, 271.

# ACKNOWLEDGEMENTS

The chase after mirages began when I chanced across Martin Swayne's *In Mesopotamia* in a house in Portugal. That got me hooked and I'm grateful to George McCall for facilitating my serendipitous encounter with the mysteries of refraction.

Different parts of the material that forms this book were presented at the Department of Art History at University College London, the Karl Jaspers Centre at the University of Heidelberg, the National College of Art in Lahore, the National College of Art in Rawalpindi, Wellesley College, the Departments of Anthropology at UCL and SOAS, and a meeting of the Swiss Graduate Program in Anthropology held in the memorable setting of Villa Garbald in Castasegna. I'm grateful to those who invited me and to all those in the audiences from whose critiques and suggestions I hope I have learned.

For critical comments and engagement I thank Farida Batool, Shaila Bhatti, Christiane Brosius, Timothy Cooper, Kate Elizabeth Creasey, Faisal Devji, Natasha Eaton, Richard Fardon, Keiko Homewood, Kostantinos Kalantzis, Shruti Kapila, Jonathan Lamb, Omar W. Nasim, Liza Oliver, Parimal G. Patil, Simon Schaffer, A. Mark Smith, Patsy Spyer, Nadeem Omar Tarar, Sarah Walpole and Yunchang Yang.

For images I'm grateful to helpful staff at the Alaska State Library Historical Collections, Biblioteca Casanatense in Rome, Museum of Fine Arts, Boston, National Portrait Gallery, Royal Anthropological Institute, Victoria and Albert Museum and the Walters Art Museum, Baltimore.

# PHOTO ACKNOWLEDGEMENTS

The author and publishers wish to express their thanks to the below sources of illustrative material and/or permission to reproduce it:

Photo Alaska State Library, Juneau: 27; from Wilhelm Bölsche, *Die Wunder de Natur* (Germany, 1913): 5; Asiatic Lithograph Press: 1, 2; photos author: 12, 13, 16, 22, 23, 35, 40, 41; collection of the author: 19; from Alexander Badlam, *Wonders of Alaska* (San Francisco, CA, 1890)/photo Alaskan Historical Society: 28; from F. W. Bain, *Bubbles of the Foam* (London, 1912): 9; photos Boston Museum of Fine Art: 20, 24; photo S. S. Brijbasi (Delhi, 2000): 36; from Royal Anthropological Institute, London: 33; Calcutta Art Studio: 8; photo Casanatense Library, Rome: 4; from *Cornhill Magazine* (London, 1883): 38; from Richard Davey, *The Sultan and His Subjects* (London, 1907): 39; from Camille Flammarion, *L'Atmosphère* (Paris, 1873): 14, 42; from *The Graphic* (London, 1885): 26; from G. Hartwig, *The Aerial World* (New York, 1874): 10; from *Illustrated London News*: 18 (1888), 31, 32 (1909); from *Kaylan* journal (1950): 7; from Joseph Meyer, *Grosses Konversations-Lexikon* (Germany, 1839): 34; from W. Haig Miller, *Mirage of Life* (New York, 1890): 17; photo National Portrait Gallery, London: 29; Friedrich Perlberg: 6; from Elisee Reclus, *The Universal Geography: The Earth and Its Inhabitants* (London, 1885): 25; photo Royal Collection Trust, London: 30; © Victoria and Albert Museum, London: 21; photo The Walters Art Museum, Baltimore: 15; from Kate Douglas Wiggin and Nora A. Smith, eds, *The Arabian Nights, Their Best Known Tales* (New York, 1937): 37.

# INDEX

Page numbers in *italics* refer to illustrations